Living OCEAN

LAB MANUAL

FOURTH EDITION

MW00435996

Kendall Hunt
publishing company

Michelle L. Hardee

Claudia Benitez-Nelson

Illustrations by Rachel E. G. Kalisperis

Contributors Tammi L. Richardson & Emily G. Baumann

Cover image: Shutterstock, Inc.

Kendall Hunt
publishing company

www.kendallhunt.com
Send all inquiries to:
4050 Westmark Drive
Dubuque, IA 52004-1840

Copyright © 2005, 2008, 2010, 2015 by Michelle L. Hardee and Claudia Benitez-Nelson

ISBN 978-1-7924-0771-0

Previous editions entitled *Introduction to Biological Oceanography Lab Manual*

Kendall Hunt Publishing Company has the exclusive rights to reproduce this work,
to prepare derivative works from this work, to publicly distribute this work,
to publicly perform this work and to publicly display this work.

All rights reserved. No part of this publication may be reproduced,
stored in a retrieval system, or transmitted, in any form or by any
means, electronic, mechanical, photocopying, recording, or otherwise,
without the prior written permission of Kendall Hunt Publishing Company.

Published in the United States of America

CONTENTS

LABORATORY

POLICIES AND PROCEDURES

1. ATTENDANCE

▶ Attendance in labs is *mandatory*. Part of your lab grade will include attendance points.

▶ If you anticipate an absence, you must inform your instructor ahead of time so we can arrange for you to attend another lab section. You *cannot* show up to another lab section without prior approval. The lab instructor will not allow you into lab.

▶ If you do miss a lab, it is *your* responsibility to find out what you missed. You will still be responsible for that material on future quizzes.

▶ Be on time to lab! Instructions and advice are given at the beginning of the period and will not be repeated.

▶ Do not expect the labs to end early—these labs are designed to run for the full three hours. You've already planned your schedule to be here for those hours, so make the most of it.

2. PREPARATION

▶ It is vital that you **read each lab assignment in full prior to attending the laboratory**. This point cannot be stressed enough: to succeed in this course, you MUST read the material beforehand. Don't worry if you don't understand everything—at least be familiar with the material so that you can work through the lab more efficiently.

3. GRADES

▶ Your grade is determined solely by *your* work and performance. Good grades are not given, they are earned. Take advantage of your instructor's office hours to get help if you are struggling, but do not wait until the last week of lab to ask for help!

4. BEHAVIOR

▶ There will be **ZERO** tolerance for cheating or plagiarism. The University's policy is clearly defined on these matters — review your student handbook. Those who feel the need to cheat will be subject to disciplinary action. This is a University, and cheating is a serious offense. You are expected to do your own work on quizzes and assignments and trust yourself to do your best.

▶ **Eating, drinking, smoking, chewing gum, and personal grooming are not allowed in the lab.** No exceptions.

▶ **Use of cell phones, pagers, all electronic devices and other distractions are not allowed.** No exceptions. They are considered a disruption and must be turned off before entering lab.

▶ No horseplay is allowed in the labs. Be respectful of other students as well as their property.

5. EQUIPMENT AND CLEAN-UP

▶ The equipment used in the laboratory serves more than one laboratory section. Report any failure or breakage so that immediate repair or replacement can be made.

▶ There are several laboratories meeting in this room all week, and it is extremely important that each laboratory cleans up after itself. You are expected to clean your work area and any equipment that you have used when you are finished. If you find a dirty lab when you come to class, report it to your instructor. *You* must leave a clean lab when you are finished.

▶ All waste should be disposed of in the proper place. Desktops should be cleaned off with a sponge and paper towels. Paper and other recyclable products should be placed in the recycling bin. Broken glass and other sharp material should be disposed of in the glass waste container. Your instructor will familiarize you with proper clean-up procedures.

6. SAFETY

▶ Some laboratory exercises may require the use of chemicals. By law, you have the "right-to-know" what chemicals are used and how exposure to these chemicals could affect your health. A copy of the Material Safety Data Sheet (MSDS) for each chemical used is available to you upon request. Your laboratory instructor will explain all the necessary precautions you need to take to protect yourself. These may include wearing goggles, aprons and/or full-face shields (these will be provided).

▶ Come to lab appropriately dressed. Any dress that is determined by your instructor to be disruptive to others will not be tolerated, and you may be asked to leave the lab if it becomes distracting. You should not wear baggy clothing, such as baggy shirtsleeves, which could get caught on (or dipped in) part of your experiment.

► **Do not wear sandals, flip-flops or any open-toed shoes to lab.** Chemicals and glassware are used in the lab room and could pose a danger to yourself and others. You will be asked to go home and change shoes if you are wearing sandals, and points will be deducted from your participation and attendance grade. This requirement will be strictly enforced in all lab sections.

IN THE EVENT OF AN EMERGENCY

► **Inform your instructor immediately.**
► **Take necessary action to prevent yourself and others from harm.**
► **Safety showers and eye wash stations** are located in each laboratory.
► **Fire extinguishers** are located near the exits.
► **First aid kits** are located at the front of each classroom.

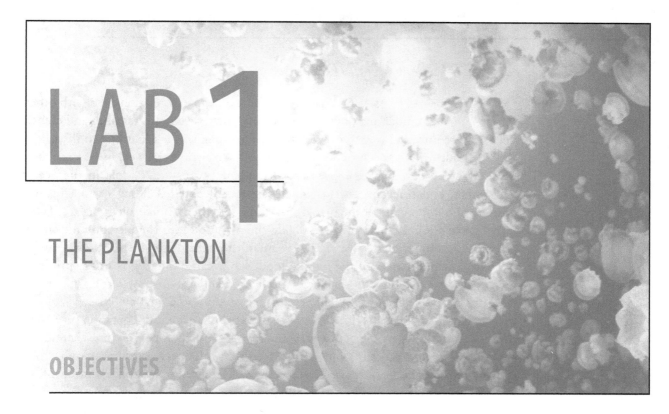

LAB 1

THE PLANKTON

OBJECTIVES

► Identify the different types of plankton present in the oceans and describe their features.
► Learn about features of organisms that relate to their adaptations for life in the plankton.
► Gain an appreciation for the great diversity of organisms represented in the plankton.

POINTS TO PONDER

► Why are there so many species of planktonic organisms if the ocean is such a homogeneous habitat (… or is it)?
► What is the benefit of having motile larval stages if you are a benthic organism?

TERMS TO KNOW

asexual reproduction	foraminifera	microzooplankton	primary production
auxospore	harmful algal bloom	nanoplankton	pseudopodia
benthos	holoplankton	nauplius	radiolarian
bioluminescence	ichthyoplankton	nekton	sexual reproduction
bloom	megalops	net plankton	test
coccolith	meroplankton	nitrogen fixation	theca
coccolithophore	mesozooplankton	phytoplankton	trochophore
cyanobacteria	microalgae	picoplankton	veliger
diatom	microbial loop	plankton	zoea
dinoflagellate	microorganism	primary producer	zooplankton

WHAT ARE PLANKTON?

All organisms can be classified based on their evolutionary relationships with other groups, such as cnidarians (jellyfish and corals), the crustaceans (shrimp and crabs), etc. A more informal way to categorize marine organisms is according to their life-style. **Plankton** are those organisms that drift or swim very weakly and are thus dependent on currents for movement. The plankton play a key role in the ocean ecosystem, and **microorganisms**, microscopic single-celled living organisms, are dominant components of the plankton. Much work in marine ecology has centered around the study of plankton. We divide the plankton into the **phytoplankton**, or "photosynthetic" plankton, and the **zooplankton**, the "animal" plankton. Phyto-plankton form the base of the marine food web because they are **primary produc-ers**, organisms that directly absorb solar radiation and use it to convert gaseous CO_2 to carbohydrates and other plant tissues. In estuaries and shallow coastal regions, benthic microalgae, marsh grasses, macroalgae or other aquatic vegetation may con-tribute to the total **primary production.** However, in the open ocean, where light does not penetrate to the bottom, most primary production results mainly from phytoplankton photosynthesis. (Refer to Lab 3: Primary Production, for more infor-mation on this subject.)

Zooplankton have many different roles in the ecosystem. Many are herbivores that feed directly on the phytoplankton, but several groups are carnivores that eat other zooplankton. In addition to categorizing organisms based on their feeding strategy (i.e., herbivore, carnivore, etc.), zooplankton can also be classified based on where the adult stage of their life cycle spends its time living. **Holoplanktonic** organisms are zooplankton that spend their entire life cycle as part of the plankton, whereas **mero-planktonic** organisms spend only their larval stages in the plankton. Some mero-plankton (such as the larvae of oysters, clams, and crabs) become benthic as adults, while others (jellies, fish larvae) remain in the open ocean realm. **Nektonic** organ-isms are typically larger animals that possess muscles for movement, and include active swimmers such as fishes, squid, and marine mammals. These organisms always remain in the water column and go through all their life stages there. The **benthos** includes all organisms attached to or capable of moving on or beneath the sea bot-tom. A planktonic larval stage can be beneficial for benthic organisms, because it provides a mechanism for the organism to move to different regions before per-manently settling to the seafloor as an adult. Meroplankton that are derived from benthic organisms are more common in shallow coastal areas, where larvae can easily metamorphose and settle. The open ocean zooplankton community is not surpris-ingly dominated by holoplankton.

Plankton can be captured by nets, which are generally made of silk or nylon. Such nets usually have mesh sizes of between 200 and 2,000 micrometers. In quantitative studies, a closing net is used so that a given depth can be sampled without contami-nation while the net is being lowered and raised. Zooplankton that are captured by a 200 μm net, but that are less than 2,000 μm in size, are called **mesozooplankton.** Zooplankton smaller than 200 μm but larger than 20 μm, usually single-celled heterotrophic ciliates or flagellates, are referred to as **microzooplankton**. This size category may also include the larger of the phytoplankton species; in that case they are called microphytoplankton. The **nanoplankton** (usually defined as 2–20 μm in size) or **picoplankton** (0.2–2 μm in size) are almost all phytoplankton, heterotrophic bacteria, or archaea and are collected by passing water through a fine mesh filter on board a ship.

THE PHYTOPLANKTON

As stated earlier, phytoplankton refers to the microscopic photosynthetic organisms found in aquatic environments. *Phyto* means "plant," and these organisms are often called the **microalgae** due to their small sizes. Some of the major phytoplankton groups are diatoms, dinoflagellates, coccolithophores, and cyanobacteria. When environmental conditions (temperature, light, nutrients) are at an optimum, populations of these organisms can increase rapidly; this is called a phytoplankton **bloom**. A bloom is typically defined as a high concentration of phytoplankton in an area that often produces discoloration of the water.

1. Diatoms

Among the most easily identified groups of microalgae are the **diatoms** (Figure 1.1 A–C). Though extremely small, individual cells often have long spines, and may be joined in a chain several millimeters long. One distinguishing feature of the diatoms is that their cell walls, sometimes called "valves" or "frustules," are made of silica (glass). Each diatom has a pair of valves that fit into each other like the two parts of a Petri dish.

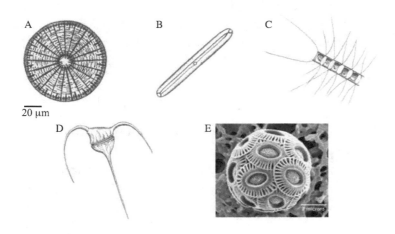

FIGURE 1.1

Types of phytoplankton. (A) Top view of a centric diatom, *Coscinodiscus*. (B) The common pennate diatom, *Amphipleura*. (C) *Chaetoceros*, a chain-forming diatom. (D) *Ceratium*, a common dinoflagellate. (E) The ubiquitous coccolithophore, *Emiliania huxleyi*. Use 20 μm scale bar for images A–D. Image 1.1.E Courtesy of Markus Geisen, Institute for Marine Resources.

Image 1.1A-D: Source: Rachel E. G. Kalisperis

During **asexual reproduction**, two new valves form inside the old ones (Figure 1.2). The two new cells formed by division are unequal in size, because one is the size of the original cell, the other smaller. This process of division is repeated until the daughter cells reach a specific minimum size. At this point, **sexual reproduction** is initiated. Valves are discarded, cells release eggs and sperm which fuse to form a zygote, which then further develops into an **auxospore**, a reproductive cell that doubles or triples its size before forming a new set of valves. After death, the valves sink to the bottom where their deposits form diatomaceous ooze, a characteristic feature of the seafloor in colder regions and upwelling zones where diatoms are common.

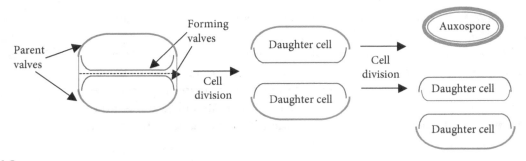

FIGURE 1.2

Cell division in diatoms requires one new valve formed for each daughter cell. Note that one daughter cell is the same size as the original parent cell and one is smaller. An auxospore forms after a specific minimum size limit is reached.

Source: Michelle L. Hardee

2. Dinoflagellates

Another very important group of single-celled phytoplankton is the **dinoflagellates** (Figure 1.1D). Most dinoflagellates have a structure called a **theca**, a cell wall composed of cellulose, and two flagellae that help in locomotion. Dinoflagellates thrive in stratified water because they can migrate vertically for light and nutrients. Because of this, in certain areas and seasons dinoflagellates may dominate the phytoplankton community. Many dinoflagellates are noted for their production of light, a phenomenon known as **bioluminescence**. This effect can be seen at night in the wake of a ship, where turbulence disturbs cells, causing them to emit light, and hence, makes the water "glow."

Certain dinoflagellate genera, such as *Alexandrium* and *Karenia*, have a notorious reputation because they can produce toxins. A **harmful algal bloom** (HAB) develops when a toxic species dramatically increases in abundance under favorable conditions. In some circumstances, the rapid production of dinoflagellates results in cell concentrations of 200,000 to 500,000 per liter. Many of these algae produce toxins such as saxitoxin and other nerve toxins, some of which are 10,000 times more deadly than cyanide. When the dinoflagellates occur in such dense concentrations, the toxin may kill ocean life over a large associated area. During feeding, bivalve mollusks filter and swallow these algae, and the toxin bioaccumulates. These tainted clams, oysters, and mussels can be dangerous and even fatal to humans.

3. Coccolithophores

Coccolithophores are unicellular phytoplankton which form part of the nano-plankton, with most species smaller than 20 µm. Their flagellated, spherical cells are covered with doughnut-shaped structures called **coccoliths** that are made of calcium carbonate (Figure 1.1E). The shape and arrangement of the coccoliths can be used to identify different species. When these coccoliths accumulate in the sediments, they form chalk. The most widespread species is *Emiliania huxleyi*, which is present in all oceans except polar seas. It can form enormous blooms, including one measured in the North Atlantic Ocean that covered an area roughly the size of Great Britain. Coccolithophores are important players in the global carbon cycle.

4. Cyanobacteria

The **cyanobacteria**, or "blue-green algae" as they used to be called, have a prokaryotic cell structure and are bacteria, not algae. The genera *Prochlorococcus* and *Synechococcus* were virtually unknown in oceanic regions in the late 1970s and 1980s. They were discovered with the development of epifluorescence microscopy and flow cytometry, instrumentation that is able to distinguish these picoplanktonic autotrophs from other small, heterotrophic bacteria by their fluorescent (photosynthetic) pigments. Scientists discovered that these organisms are perhaps the most numerous phytoplankton in the open ocean, numbering up to 1,000,000 cells/mL in some regions. Another common cyanobacterial genus is *Trichodesmium*, which forms large mats in surface waters. The cyanobacteria have been shown to account for a large fraction of marine primary production, and many species convert N_2 (a process called **nitrogen fixation**) into ammonium, an essential nutrient for primary producers.

The picoplanktonic cyanobacteria (*Synechococcus* and *Prochlorococcus*) are key players in a cycle of matter and energy within marine ecosystems called the microbial loop. Also involved in the microbial loop are heterotrophic bacteria, which consume dissolved organic matter (DOM) that cannot be directly ingested by larger organisms. DOM is a diverse mixture of small molecules that are derived from living and decomposing organisms, and it is the largest reservoir of organic matter in the ocean. Marine bacteria are consumed by microzooplankton, which are in turn fed upon by mesozooplankton, linking the microbial loop to the traditional marine food web, which is comprised of large phytoplankton, large grazers, and their predators.

THE ZOOPLANKTON

The zooplankton incorporate a diverse array of organisms, from the microscopic single-celled protists to some ctenophores and cnidarians that can stretch for meters in length. The major zooplankton groups are shown in the following diagrams. The groups labeled as larvae are meroplanktonic, and the rest are holoplanktonic.

1. Foraminifera

All organisms in the Kingdom Protista are single-celled, and the **foraminifera** (Phylum Foraminifera or Sarcomastigophora) are an intensively studied group, particularly for climate research. Foraminifera possess **pseudopodia**—extensions of the cell cytoplasm that can extend and retract for feeding and other functions. These organisms (called 'forams' for short) often have a multichambered **test** (the shell of a single-celled organism) that is composed of $CaCO_3$ (Figure 1.3A). There are both planktonic and benthic foraminiferal groups; the tests of planktonic forms have long spines that aid in flotation. The pseudopodia protrude through the pores in the test and along the spines to capture food in a mucus-like web of material, which is then retracted into the cell for feeding. The shells of planktonic forams eventually sink to the sea bottom, where they form foraminiferal ooze, a type of calcareous ooze consisting mostly of foram tests.

2. Radiolaria

Radiolarians are single-celled planktonic marine protozoans belonging to the Kingdom Protista. They secrete elaborate tests made of silica and other materials, and form shapes such as spheres, crowns, or stars with numerous radiating spines (Figure 1.3B–C). Thin pseudopodia capture food in a manner similar to foraminifera. Radiolarians are found in open ocean habitats, and when they die they settle to the seafloor and form radiolarian ooze.

FIGURE 1.3
Some representative groups of the single-celled zooplankton. (A) A planktonic foraminiferan. Note the chambers and aperture. (B) A spherical radiolarian with portions of its outer sphere broken. (C) A vase-shaped radiolarian. Scale bar is approximate. Images A and B by Michelle L. Hardee.

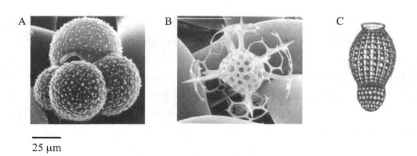

25 µm

Image C: Source: Rachel E. G. Kalisperis

3. Larval groups

Several groups of organisms have larval stages that are found in the zooplankton (refer to Figure 1.4). The majority of these groups include several types of crustacean larvae. The **zoea** and **megalops** are the larval stages for crabs (Phylum Arthropoda). The **nauplius** is a larval stage for barnacles, ostracods, copepods, and other smaller

crustaceans. As the organisms pass through these different larval stages during development, the larvae become progressively more adult-like in appearance. Polychaete
worms (Phylum Annelida) have a larval stage called a **trochophore,** which looks
like a top with rows of cilia. Some mollusks also have a trochophore larva. Other
mollusks, like the gastropods, have a characteristic larval stage called a **veliger**, which
looks somewhat like a butterfly with a shell. Echinoderm classes have unusual and
characteristic larvae, including the ophiopluteus (brittle star) and brachiolaria (sea
star).

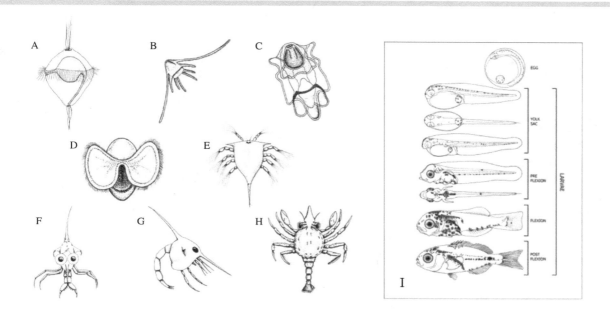

FIGURE 1.4

Meroplanktonic larvae of benthic invertebrates (left). (A) Polychaete trochophore. (B) Brittle star ophiopluteus. (C) Sea star brachiolaria larva. (D) Snail veliger.
(E) Barnacle nauplius. (F) Crab zoea, front view. (G) Crab zoea, side view. (H) Crab megalopa. (I) Planktonic stages of a fish. Image I from Brownell, C. L., 1979, *Stages in
the Early Development of 40 Marine Fish Species with Pelagic Eggs from the Cape of Good Hope*, J. L. B. Smith Institute of Ichthyology, Rhodes University.

Image A-H: Source: Rachel E. G. Kalisperis

A vital component of total zooplankton is **ichthyoplankton**, the eggs and larvae
of fish (Figure 1.4I). Once hatched, fish larvae have almost no swimming ability
and feed on smaller plankton and are, in turn, prey themselves for larger animals.
Monitoring egg and larval counts are a good way to monitor trends in population
abundance of the adults and help indicate a healthy or stressed ecosystem.

4. Other groups

Several other animal phyla are represented in the zooplankton. One of the most
abundant groups is the chaetognaths, or arrow worms (Figure 1.5A). These are
extremely important predators in the zooplankton, feeding mostly on copepods as

well as whatever else they can capture. Another very abundant group of zooplankton are the crustaceans (Phylum Arthropoda), which include copepods, ostracods, and amphipods (Figure 1.5B–E). Most zooplankton are crustaceans, and the most abundant crustaceans in the zooplankton are the copepods. In South Carolina estuaries, copepods comprise about 70% of all the zooplankton and often reach densities of 50,000 animals per cubic meter of water! Jellyfishes and hydrozoans such as the man-o'-war are also classified as zooplankton, as they are carnivorous but unable to actively swim against currents. There are numerous other groups that populate the zooplankton that will not be mentioned here.

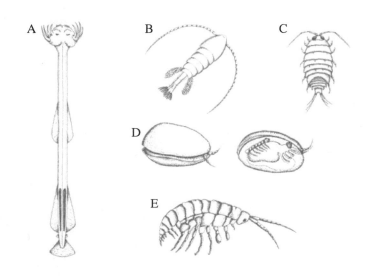

FIGURE 1.5

Some of the major organism phyla that are classified as zooplankton. (A) The predatory chaetognath. B–E: Arthropods. (B) The copepod *Cyclops*. (C) A typical isopod. (D) An ostracod. The left illustration shows the typical view, with its clam-like carapace valves covering the crustacean. The right illustration shows a simplified diagram of the organism with one of the carapace valves removed. (E) *Gammarus*, an amphipod.

Source: Rachel E. G. Kalisperis

EXERCISE 1.1
INVESTIGATING THE PHYTOPLANKTON USING THE *PHYTOPIA* PROGRAM

Materials required: Computer access with *Phytopia* program and headphones

Directions

In this exercise you will be using a computer program called *Phytopia* to examine the different groups of phytoplankton. Follow the instructions and answer the questions below. Your instructor will show you how to navigate the program.

1. First, familiarize yourself with the layout of the program design. There are three main modules: **Phyto Files, Phyto Factors,** and **Special Topics.** You can either choose from the menu at the bottom left or move the microscope until the desired module appears. Note the menu button and other navigational icons.

2. Select the **Phyto Factors** module. Note that you can pause, rewind, and restart sections of the module. As you watch and listen to the movie, take notes on the information and answer the following questions.

    a. What are some of the large-scale processes (oceanic, atmospheric, etc.) that affect the ocean's physical environment and thus can affect phytoplankton growth? _____

    _____

    _____

    b. What are the three factors critical to phytoplankton survival? _____

    _____

    _____

    c. Which two of these are actually limiting to phytoplankton growth? _____

    _____

    _____

    d. What factors affect the depth of the photic zone? _____

    _____

e. In general, the water column can be described as either well-mixed or stratified. How does a well-mixed water column contribute to phytoplankton growth and blooms?

_____

_____

f. How does a stratified water column limit phytoplankton growth?

_____

_____

3. Start the **Bloom Activation Tool** in the **Phyto Factors** module. Listen to the introduction and follow the instructions. Your goal is to determine the optimal conditions for a phytoplankton bloom in the Gulf of Maine, taking into consideration photosynthetically active radiation (PAR), sea surface temperature (SST), and wind conditions. Once you have made your selection of these three factors, click on the "Activate" button to see if a bloom is produced. Remember, this is the Gulf of Maine, which is subpolar in its latitude and climate. You may have to do this several times to be successful.

a. What were the conditions that activated a bloom?

PAR: _____

SST: _____

Winds: _____

b. Why do you think these particular conditions had to co-occur in order to activate a bloom? _____

_____

_____

4. Click on the Menu button and select the **Phyto Files** module. This portion of the program allows you to examine different species of phytoplankton, including important characteristics and interactive microscope views. The benefit of this exercise lies in the ability for you to view many different phytoplankton species, most of which are either too small to see using the available microscopes, are extremely difficult to obtain, or can pose a hazard to human health.

5. There are multiple ways of navigating this module. First, read the instructions in the top right window. The Mode window below it will list your chosen parameters as you select different groups. The bottom window in the blue section lists genera and species; this will change as you alter your attribute choices in the section to the left of this window. You have five attributes that you can use to search; each attribute has additional subtypes that will help to narrow a search for particular types of phytoplankton. As you narrow your attribute list, note that the species list becomes shorter as well. Select different attributes and watch the bottom two windows change as your attributes vary. *Important*: if only two attributes are specified in a question, make sure the other attributes have "All Selected," or you won't reach the correct answer.

6. When you click on a species name in the bottom right window, the species will appear in the "microscope," and the organism's attributes will appear in the top window. If an attribute is underlined, when you click on it there is additional information that appears in the middle window. Note the different types of microscope views that you can use to examine the organisms. Not all organisms will have every view as it depends on the images available. Familiarize yourself with the magnification and navigation buttons for the microscope view.

7. In **Attributes**, select **Class: Diatoms**, then **Morphotype: Chains**.

   a. How many genera were listed? _____ Give an explanation for why you think there are so many chain diatom species listed. _____

   _____

   b. Choose one of the genera and sketch it below (be sure to provide its genus name):

   c. What is the composition of its cell wall? _____

   d. What is its general distribution in the oceans? _____

8. Select **Class: Dinoflagellate, Shape: Triangular**, and **Morphotype: Solitary**.

   a. What genera are listed? _____

   b. Are any of these dinoflagellate genera harmful? _____

   c. *Ceratium* is a common dinoflagellate. Click on the genus name and draw what you see below:

d. What is the composition of this species' cell wall? _____

e. What interesting physiological feature of this organism is described in **"Other Facts?"**

_____

9. Select **Class: Dinoflagellate, Harmful,** and make sure other attributes have **"All Selected."**

   a. Click on *Karenia brevis*. In what ocean conditions do they live?

   _____

   b. Click on *Gambierdiscus toxicus*. What is their distribution and in what type of ocean conditions?

   _____

   _____

   c. Why do you think particular harmful dinoflagellate species only occurs in specific geographic
   areas or under certain oceanic conditions? _____

   _____

   _____

   d. Click on *Alexandrium tamarense*. A common rule of thumb for eating shellfish is described under
   Other Facts. What is it? _____

   _____

   e. How do you think climate change will affect this often-invoked rule of thumb? _____

   _____

10. Select **Class: Diatom,** and **Harmful.**

    a. How many species are listed? _____

    b. Click on *Chaetoceros socialis* and draw it below:

c. How is this species harmful to fishes? _____

d. Click on *Nitzschia sp.* and sketch it below:

e. In what way is *Nitzschia* harmful?_____

f. *Nitzschia* contains storage products including chrysolaminarin and oils. Why would a diatom need to store such products? _____

_____

11. In **Class,** select **Other,** and for all other attributes, select **All Selected.**

a. Click on *Trichodesmium* species and draw it below.

b. Select "**Extra Images.**" What do these images indicate about the lifestyle and habitat of *Trichodesmium*? _____

_____

_____

c. What group (listed in **Class**) does this organism belong to? _____

d. What important role does *Trichodesmium* play in the oceans (click on **Other Facts**)?

_____

_____

e. Click on *Phaeocystis* sp. This species is harmful but it also produces a feature that you may often see on beaches. What is it? (Select **Harmful** attribute.)

_____

_____

f. How can *Phaeocystis* affect the weather? (Select **Other Facts** attribute.) _____

_____

_____

EXERCISE 1.2
INVESTIGATING THE ZOOPLANKTON

Materials required: Dissection and compound microscopes, prepared microscope slides, microscope slides and cover slips for wet mounts (if material is available), identification books

Directions

In this Exercise you will examine and become familiar with the different groups of zooplankton. Answer the questions associated with the slides set out for you. Review **Appendix A** if you need a refresher on how to use a microscope.

1. Foraminifera

a. Draw two different foraminifera that you see in the space below (look along the edges of the coverslip if you can't find them):

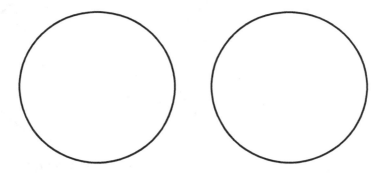

b. Is this a single- or multi-celled organism? _____

c. What do the chambers represent? _____

d. What is the test of this organism composed of? _____

2. Radiolaria

   a. Draw two different radiolarians that you see in the space below:

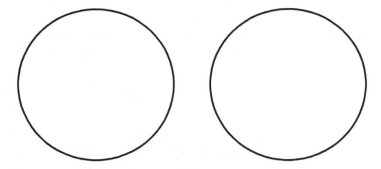

   b. What is the test of this organism composed of? _____

3. Crab larvae

   a. Draw the two major larval stages: zoea and megalops in the spaces below:

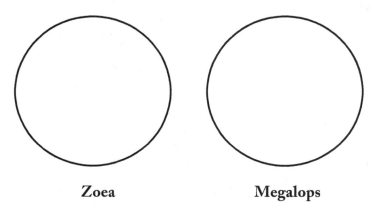

   Zoea   **Megalops**

   b. What is the advantage of the long spine on the zoea? _____

   _____

   _____

4. Oyster larvae

    a. Draw the larval stages in the space below.

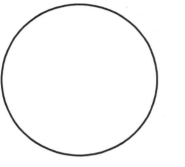

           Trochophore   **Veliger**

    b. Describe the differences between the two stages.

_____

_____

5. Chaetognath

    a. Draw the chaetognath in the space provided:

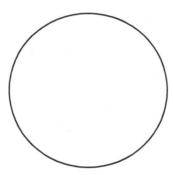

    b. What is the function of the long, sharp spines around the chaetognath's mouth?

_____

    c. Though they are less than a centimeter in length, chaetognaths are highly predatory. What do you think is the major prey item of chaetognaths and why? _____

_____

_____

6. Ostracod

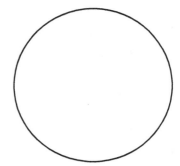

    a. Draw the ostracod in the space provided:

    b. To what phylum does this organism belong?

        _____

    c. What features might have misled you when you were attempting to identify the ostracod's phylum? _____

        _____

7. Copepod

    a. Copepods are the most numerous zooplankton in the pelagic realm. Why do you think this is so? _____

        _____

    b. Draw a male vs. a female copepod in the space below.

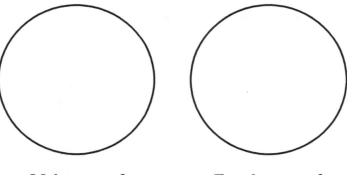

Male copepod   **Female copepod**

    c. What features do you use to distinguish the sexes? _____

        _____

8. Ichthyoplankton

a. Draw a larval fish in the space provided:

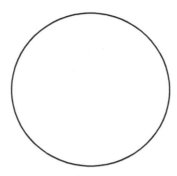

b. If this fish grows 0.3 mm per day, how old is this fish?

_____

9. Holoplankton vs. Meroplankton

From the plankton observed in lab today, list three examples for the following types of plankton:

Holoplankton	Meroplankton

LAB 2

METHODS OF CHLOROPHYLL ANALYSIS

OBJECTIVES

- ► Quantify the concentration of chlorophyll α in a seawater sample using fluorometry.
- ► Understand the underlying principles of the chlorophyll fluorometry approach.
- ► Learn how satellite oceanography provides estimates of chlorophyll.
- ► Learn how to interpret satellite images and draw conclusions from the data.

POINTS TO PONDER

- ► Why are scientists interested in the measurement of chlorophyll α?
- ► Why do coastal regions have higher chlorophyll concentrations than open ocean regions?
- ► What are the advantages/disadvantages of ship and satellite-based measurements?
- ► Why are there so many different types of satellites when they all measure similar wavelength bands?

TERMS TO KNOW

active sensors	CZCS	*in situ*	primary production
altimetry	fluorescence	MODIS	SeaWiFS
bands	fluorometer	ocean color	synoptic
biomass	GIS	passive sensors	
chlorophyll α	ground-truth	photosynthesis	

WHAT IS CHLOROPHYLL α?

As you have learned, the world's oceans teem with life, with microscopic phytoplankton occupying the sunlit zone. Phytoplankton are fundamental to the oceanic food web as they convert the energy in sunlight and carbon dioxide (CO_2) into oxygen and food for other organisms, in a process called **photosynthesis** (see Lab 3 for more information). All photosynthetic organisms, whether terrestrial or aquatic, contain the primary photosynthetic pigment **chlorophyll α** (Figure 2.1). This pigment strongly absorbs red and blue light but not green wavelengths, causing this pigment to appear green to our eyes.

Over 70% of the surface of the earth is covered in water. Phytoplankton are responsible for almost half of Earth's **primary production**, the amount of CO_2 fixed via photosynthesis over time. Yet phytoplankton **biomass** (the total weight of an organism in a given volume) accounts for *less* than 1% of global plant biomass. As you can imagine, there is tremendous scientific interest in measuring phytoplankton abundance. Because all phytoplankton contain chlorophyll α, this pigment can be used as a "proxy" for phytoplankton biomass.

CHLOROPHYLL α ANALYSIS BY FLUOROMETRY

How do we measure chlorophyll α? **Fluorescence** is the phenomenon of some compounds (such as pigments) absorbing specific wavelengths of light and almost instantaneously re-emitting this light energy. Because there is a slight loss of energy (due to absorption rather than re-emission), the emitted light is of a longer wavelength than what was absorbed (Figure 2.2). Thus, chlorophyll α naturally absorbs blue light and emits, or fluoresces, red light. **Fluorometers** detect chlorophyll α by

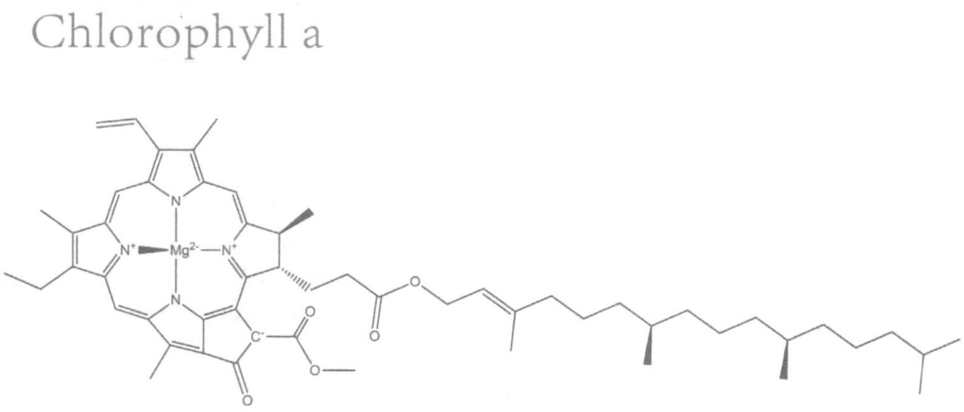

FIGURE 2.1

Molecular structure of chlorophyll. C = Carbon, H = Hydrogen, N = Nitrogen, O = Oxygen, Mg = Magnesium. Lines represent Carbon—Carbon bonds. R = side chain group. In chlorophyll α, the R group is a methane group (CH_3).

Chlorophyll a

© Raimundo79/Shutterstock.com

transmitting an excitation beam of blue light and then detecting the red light (~685 nm) fluoresced by phytoplankton cells or extracted chlorophyll in a sample. Generally, this fluorescence is directly proportional to the concentration of the material in question.

Chlorophyll α fluorescence is preferred by scientists for quantification of phytoplankton because it is the most versatile, sensitive, and easiest way to measure the concentration of phytoplankton in water. The estimation of chlorophyll α concentration provides information on phytoplankton abundance. Since chlorophyll-containing organisms are the first step in ocean food webs, the abundance of these organisms results in cascading effects to all organisms dependent on these phytoplankton as either a direct or indirect food source. Chlorophyll α concentration is one of the key indices in monitoring the health of an environment.

While measuring chlorophyll α is faster and easier than counting phytoplankton cells by microscopy, it provides only a "bulk" estimate of phytoplankton biomass. It does not provide information about species composition ("who" is in the sample). Also, the relationship between fluorescence and the actual concentration of chlorophyll α can be altered by varying environmental parameters from optimal conditions (Figure 2.3). Cell physiology and morphology, variations in light through the cell's life cycle, and the presence of interfering compounds can alter the amount of light fluoresced by chlorophyll in cells. Interfering compounds can include other plant pigments, degradation products, dissolved organic matter, and turbidity (cloudiness of the water). Thus, fluorescence data give an estimate of chlorophyll concentrations which are, in turn, an estimate of cell abundances in the ecosystem.

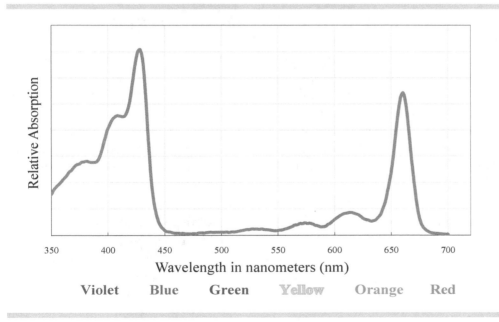

FIGURE 2.2

The absorption spectrum of chlorophyll α. The absorbance of visible light by chlorophyll *a* is measured in a spectrophotometer as a function of wavelength. The absorption maximum is about 460 nm.

Source: Michelle L. Hardee.

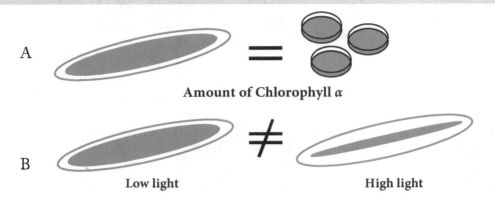

FIGURE 2.3
The comparison of chlorophyll α concentration in the cell and how (A) different cells can contain equal amounts of chlorophyll, whether it's one large cell or multiple smaller cells; or (B) depending on the light environment, cells can contain varying amounts of chlorophyll, even though they are the same cell type.

Source: Emily G. Baumann

Chlorophyll measurements via satellites are also used to directly monitor phytoplankton populations. As chlorophyll concentrations increase and therefore blue wavelengths of light are removed from the water, the water will appear increasingly green to our eye. Satellites carrying sensors that are sensitive to changes in this "water-leaving radiance" can detect changes in ocean color from space. Numerous satellites have been launched to examine surface ocean characteristics including chlorophyll concentrations, temperature, salinity, silt/sediment patterns, and circulation.

SATELLITES AS INSTRUMENTS FOR OBSERVATION

For many features that we want to learn about in the oceans, such as primary productivity, temperature, current patterns, etc., it is often necessary to take a step back and see the bigger picture. Conditions in the oceans can change quickly, and that fact, combined with their enormous size, means that the chances of truly characterizing the conditions in any given oceanic region are quite difficult. Ship-based oceanographers are still generally limited to sampling over a small area and short time intervals, and these data are often obtained with a great deal of difficulty.

Satellites, however, are excellent tools for looking at very large areas of the oceans over a very short time period. This is called **synoptic**, the ability to look at a large region at the same instant. The following satellite image on the internet is composed of many individual measurements that a satellite acquired as it scanned the earth from space. Type this web address into your internet browser to view the video showing a typical satellite path:

https://www.youtube.com/watch?v=d4QLDlAumOc

In this video, the individual measurements (called pixels) are spaced approximately 1 kilometer apart. This particular image represents a portion of the earth that is about 2,300 km², meaning that there are 1 million individual satellite measurements in this single image. Imagine that you are on a research ship in the waters around Tasmania (Australia), steaming east along a 2,300-kilometer cruise track at about 10 knots. When you reach the end of one row, you head north for 1 kilometer, turn around and then head west for another 2,300 kilometers. This process of working back and forth across the ocean is repeated until you have matched the coverage that you see in the satellite image. However, while it took the satellite just a little more than 1 *minute* to make those 1 million measurements, it would take you more than 10 *years* to make the same number of measurements from the ship. On board ship, you would also not be able to measure with any temporal resolution. Satellites will give you that ability with their synoptic capabilities, but they are not a substitute for actual *in situ* data. Therefore, scientists must also **ground-truth**, or sample the ocean (or land) directly, and combine these data with satellite measurements to obtain accurate and high-resolution data coverage.

Among the first satellites in the United States' entry into space were those designed to monitor weather systems and climate variations at regional and continental scales, and for mapping Earth's topography. These satellites included the Landsat series, of which Landsat 5, 7, and 8 are still in orbit and operational, and the Nimbus series, used primarily for atmospheric monitoring. The data gained from these satellites greatly improved real-time surveillance of clouds, land, and ocean temperature variations, water vapor, and meteorological phenomena, as well as tremendous improvements in topographic mapping. Each satellite has a different set of complex instruments (sensors) that measure different **bands** (see Table 2.1). Bands are wavelength intervals that

TABLE 2.1

Comparison of Landsat-8 and MODIS bands with their respective measured wavelengths and visible spectrum responses. IR = Infrared. 1 nm = 10^{-9}m, 1 μm = 10^{-6} m. Information from: http://rst.gsfc.nasa.gov/Front/tofc.html and http://modis.gsfc.nasa.gov/about/specifications.php.

Landsat-8			MODIS		
Band No.	Wavelength Interval	Spectral Response	Band No.	Wavelength Interval	Spectral Response
1	433–453 nm	Blue	8	405–420 nm	Violet-Blue
2	450–515 nm	Blue-Green	9	438–448 nm	Blue
3	525–600 nm	Green-Yellow	10	483–493 nm	Blue-Green
4	630–680 nm	Orange-Red	11	526–536 nm	Green
5	845–885 nm	Mid-IR	12	546–556 nm	Green-Yellow
6	1.56–1.66 μm	Thermal IR	13	662–672 nm	Red
7	2.10–2.30 μm	Thermal IR	14	673–683 nm	Red
8	500–680 nm	Green-Red	15	743–753 nm	Red
9	1.36–1.39 μm	Thermal IR	16	862–877 nm	Mid-IR

represent different parts of the visible, infrared, or ultraviolet energy spectrum. Sensors can be either passive or active. **Passive sensors** detect naturally reflected or radiated energy from their target (Sun, Earth, or atmosphere). Examples of such energy are thermal infrared waves or the visible spectrum emitted by the earth. **Active sensors** emit, or send out, their own electromagnetic energy and then record the energy reflected back to them. Examples of these sensors are those that use radar (microwave radiation) or lasers (visible spectrum). One advantage of active sensing systems is that, since they provide their own source of energy, they can collect data at any time of the day or night. Passive sensors must rely on receiving either naturally emitted or the sun's reflected energy from Earth's surface. A valuable application of active sensors is satellite **altimetry**, the determination of the sea surface height relative to sea level as measured from a satellite. After accounting for waves and tides, the surface of the ocean bulges outward and inward very slightly, mimicking the topography of the ocean floor. The bumps, while too small to be seen, can be measured by a radar altimeter aboard a satellite. The level of resolution is such that even slight variations in the sea level resulting from El Niño-induced warm water currents can be measured.

As you have learned, the world's oceans teem with life. Central to the marine food web are phytoplankton. The process of photosynthesis primarily depends on the phytoplankton's content of chlorophyll a, a pigment that strongly *absorbs* red and blue light and *reflects* green wavelengths. As phytoplankton concentrations increase, there is a corresponding rise in the spectral radiances emitted from the plankton at very specific wavelengths, peaking in the green portion of the visible spectrum. Upwelling masses of water containing phytoplankton take on green hues that contrast to the deep blues of ocean water with few nutrients. Satellites launched to examine ocean chlorophyll concentrations include the Coastal Zone Color Scanner (CZCS) sensor on Nimbus 7, the SeaWiFS instrument on the OrbView-2 satellite, and MODIS instrument on the Terra and Aqua satellites (1999–present).

The **Coastal Zone Color Scanner (CZCS)** was a sensor specifically developed to study ocean color properties. Subtle changes in ocean color signify various types and quantities of marine phytoplankton. It sensed colors in the visible region in four bands (refer to Table 2.1). Band 1 (blue) measured chlorophyll absorption; band 2 (green) tracked chlorophyll concentration; band 3 (yellow) was sensitive to yellow chlorophyll pigments, and band 4 reacted to aerosol (dust particle) absorption. A fifth band monitored surface vegetation and band 6 sensed sea surface temperatures. Data from these "**ocean color**" bands allowed scientists to calculate chlorophyll variability, and hence the relative abundance of marine phytoplankton: blues correspond to the lowest levels of phytoplankton and reds to the highest.

The CZCS mission revolutionized the use of satellite imagery for oceanographic research. The ocean color satellite images of Earth that have become so widely used were obtained from this satellite's measurements. The success of the CZCS mission (which lasted from 1978 to 1986) is what led NASA to develop the **SeaWiFS** program (*Sea*-viewing *Wi*de *F*ield-of-view *S*ensor), launched in August of 1997 and ended in December 2010. The purpose of SeaWiFS was to provide quantitative data on global ocean color properties. On SeaWiFS, several bands covered the blue, green, and red wavelengths of the visible spectrum and into the near infrared, yielding data that could be used to display chlorophyll variations of surface phytoplankton. Like CZCS, SeaWiFS produced regional scale images, which allowed scientists to monitor eddies, open ocean circulation patterns, and sedimentation patterns in coastal areas. Since an orbiting sensor can view every square kilometer of ocean every 48 hours, satellite-acquired ocean color data are a valuable tool for determining the abundance and patterns of ocean life on a global scale, and can be used to assess the ocean's role in the global carbon cycle.

The **MODIS** (*Mod*erate Resolution *I*maging *S*pectroradiometer) instrument is currently aboard both the Terra and the Aqua Earth Observing System (EOS) satellites. This instrument measures not only ocean color but also aerosol properties, atmospheric temperature, and water vapor. One of the major benefits of the MODIS instrument is that it provides continual, comprehensive coverage of the globe every two days. Terra's orbit around Earth is timed to pass from north to south across the equator in the morning, while Aqua passes south to north in the afternoon, thus the two instruments complement one another and optimize the coverage of data.

The most widely used application of Earth-observing data from satellites is from the correlation and interweaving of multiple types of data that when combined, become essential to policy decision-making and the application of predictive models. In addition to oceanography, numerous other disciplines are concerned with Earth observations and resources, including meteorology, climatology, coastal processes and management, land management, military applications, etc. The bulk of the data in these systems make up what has become a powerful tool in these fields: the **Geographic Information System** (GIS). Because vast amounts of spatial or geographically referenced data must be gathered, stored, analyzed in terms of their interrelations, and rapidly retrieved for day-to-day decisions, a GIS must be computerized to be used efficiently. Without computers, remote sensing from space would be next to impossible.

EXERCISE 2.1
MEASURING CHLOROPHYLL α USING FLUOROMETRY

Materials required: 3 Glass Fiber Filters (GF/C, 1.2 μm pore size), filter forceps, 3 glass vials with screw tops, 3 culture tubes, filter funnel and frit, Erlenmeyer filter flask, vacuum pump, 10-ml graduated cylinder, squirt bottle with filtered seawater, small beaker to hold culture tubes, container of pre-made seawater sample with phytoplankton, Kimwipes™, Parafilm™, access to a Turner Designs fluorometer, and 90% acetone bottle.

A. METHODS

Directions

1. Keep room lights as low as possible (turn off bright fluorescent lights).

2. Make sure your filtration equipment is assembled: filter flask, vacuum pump, funnel, frit.

3. Place a clean filter on your filter funnel as demonstrated by your instructor. Use the forceps to handle the filter—do not use your fingers.

4. Bring a clean glass beaker to obtain a sample of seawater from your instructor. Before pouring, mix the sample by gently inverting the container—do not shake. Once the sample is mixed, use the graduated markings to pour about 100 mL of seawater sample into your beaker and take this back to your group.

5. Now, using a graduated cylinder, measure out exactly 10 mL of your obtained sample (this is called an aliquot). Record this value in Table 2.2 for 'Volume of seawater filtered'.

6. Gently pour the 10 mL aliquot into the filter funnel and begin to vacuum pump. Rinse the sides of the filter funnel using the squeeze bottle to rinse any residual phytoplankton onto your filter. Continue to vacuum pump until the entire sample has been filtered. (Note: Because the squeeze bottle water is pre-filtered, it is not necessary to record this volume.)

7. Disassemble the filter funnel, and remove the filter using forceps (be careful not to disrupt the material in the center of the filter). Place the filter into a glass vial with a screw-top lid. Label your lid with a group ID and replicate number.

8. Add 5 ml of 90% acetone and make sure your filter is submerged. Record this volume in Table 2.2 under 'Volume of acetone for extraction'. Wrap your vial in aluminum foil keeping the label on the lid visible.

9. Repeat steps 5–8 for two more aliquots of your seawater, for a total of three replicate aliquots.

10. Now place your 3 vials with acetone and filters into the freezer. Record the start time of your extraction in Table 2.2.

11. Allow the pigments to extract for one hour.

12. During the extraction process, proceed to **Exercise 2.2.**

13. After the 1-hour extraction, remove the vials from the freezer and allow them to come to room temperature (you can warm them with your hands).

14. Read the chlorophyll concentration on a calibrated fluorometer following these instructions:

    a. Gently mix your sample and pour off approximately 2 ml of your first replicate (about half your sample volume or enough to fill the window inside the fluorometer) into a clean glass culture tube.

    b. Wipe the sample tube with a Kimwipe to remove any fingerprints.

    c. Insert your sample tube into the instrument and close the lid.

    d. Press "Measure Fluorescence".

    e. Follow the prompts to enter in your volume filtered and your acetone (solvent) volume. Watch for the reading and record the value of the first replicate in Table 2.2.

    f. Repeat steps a–e for your remaining replicates.

15. Finally, average your three volume-corrected concentrations to calculate the final chlorophyll α concentration (µg/L). Record your group's result on the whiteboard so that you can compare the values calculated by all groups in the class.

16. When you have finished your calculation, clean your tables, dispose of the acetone in the waste container, and rinse out all of your glassware thoroughly.

TABLE 2.2

Values for chlorophyll α samples measured by fluorometry.

	Replicate 1	**Replicate 2**	**Replicate 3**
Volume of seawater filtered (ml) [V_f]			
Volume of acetone for extraction (ml) [V_a]			
Fluorometer measurement (µg/L) [F]			
Final Chlorophyll α concentration (µg/L) [average]			

B. DATA ANALYSIS AND RESULTS

1. What is the chlorophyll α concentration of the seawater sample? _____

2. Record the values of coastal, estuarine, and open ocean chlorophyll that you found during Exercise 2.2 in the space below.

   _____

   _____

3. How did your chlorophyll concentrations (measured in the lab) compare to concentrations in other marine environments? To which marine environment are your values closest?

   _____

   _____

   _____

4. Convert the following chlorophyll concentration units: 20 mg/m^3 = _____ µg/l

C. DISCUSSION

1. Why is chlorophyll α a good estimator of total phytoplankton biomass in seawater?

   _____

   _____

2. Consider the satellite images of chlorophyll concentrations along coastlines of the U.S. you observed in Exercise 2.2. How does the concentration of chlorophyll α vary from coastal to open ocean (close to shore, to offshore)?

   _____

   _____

   _____

3. What environmental factor do you think is responsible for this observed onshore-to-offshore distribution of chlorophyll α?

   _____

   _____

4. The chlorophyll α distribution images you examined in lab show chlorophyll concentration gradients that seem to run counter to logic (or intuition), in that the lowest amounts of chlorophyll, and therefore phytoplankton, are in the equatorial and subtropical zones, particularly the subtropical gyres. Typically, such warmer regions should support greater quantities of plant life as do tropical regions on land. Explain why subtropical and equatorial zones have lower than anticipated chlorophyll concentrations.

    _____

    _____

    _____

EXERCISE 2.2
INTERPRETING SATELLITE IMAGERY

Materials needed: computer with internet access

DIRECTIONS

This lab is an exploratory-based exercise to examine satellite oceanographic images and interpret what you see. During the chlorophyll α extraction process in Exercise 2.1, you will be accessing online sources of sea surface chlorophyll data, and record representative values (or ranges of values) for chlorophyll concentrations in coastal oceans, open oceans, and estuaries. Go through the questions in order, as they take you step-by-step through the websites and interpretation.

1. Go to the following site: http://svs.gsfc.nasa.gov/. This site provides a listing of animations and images available from the SeaWiFS and MODIS instruments. There are over 6,900 images available for viewing. Click on SEARCH. Use the search bar to enter the I.D. number for each image required for the questions below. You can also search these animations by keywords.

2. Search for the movie #12777, **Our Living Planet From Space.** Watch the video and answer the following questions.

    a. Why is having 20 years of continuous data of Earth's surface so important?

    _____

    _____

    b. Why is it particularly important to be able to observe chlorophyll α from space?

    _____

    _____

3. Search for the image #4596, **20 Years of Global Biosphere.** Read the information provided. Referring to the ocean chlorophyll α concentration color scale and the information provided on this page, refer to this color scale for questions throughout this exercise.

    Red: 30–50 mg/m³ (high)

    Green: ~1.0 mg/m³ (medium)

    Purple: ~0.1 mg/m³ (low)

Now examine the animation for changes in ocean color variations over the 20-year period of hydro-sphere data and answer the following questions:

a. Which ocean areas or ecosystems (open ocean, coastal, equatorial, etc.) exhibit the highest pro-ductivity in the oceans during the majority of the 20-year time series? Where are the lowest pro-ductivity areas?

_____

_____

b. Examine the seasonal color change on the continents and compare them to large-scale ocean pro-ductivity changes. What is the relationship between terrestrial seasonal changes and large-scale ocean productivity variation?

_____

_____

c. In which ocean region(s) do you see the most noticeable fluctuations in productivity with the change of seasons?

_____

_____

What are the factors controlling this high seasonal variation in primary productivity?

_____

_____

4. The Equatorial Pacific Ocean is one of the most important ocean areas where ocean and climate change patterns related to El Niño are studied. A recent El Niño event occurred in 2015–2016, as evidenced by the variability in primary productivity in this region during the event. Use the Search to look for image #4387, **El Niño: Disrupting the Marine Food Web,** and then answer the following questions:

a. Describe the pattern of ocean color changes related to primary production with time across the equatorial Pacific during the El Niño event: _____

_____

_____

b. A major oceanographic physical feature in the Eastern Equatorial Pacific associated with South America is significantly influenced by El Niño perturbations. What is it? _____

_____

5. Satellite oceanographers can obtain quantitative data from satellite images and analyze for changes. A tool called the Geospatial Interactive Online Visualization and Analysis Infrastructure (Giovanni for short) is a web-based application developed by the Goddard Earth Sciences Data and Information Services Center (GES DISC). This system provides easy and open access to vast amounts of NASA remote sensing data for scientific exploration and discovery without having to download the data.

Knowing that we experienced a recent El Niño event, we can extract and analyze data from the satellite images. Let's look at the average monthly chlorophyll α concentrations for the years before, during and after the El Niño event. Open a new window on your browser and go to the Giovanni website: https://giovanni.gsfc.nasa.gov/giovanni/ or search for "Giovanni" and "NASA". Click on the link to take you to the Giovanni program.

On this page, choose the following parameters:

► Select Plot: Time Series: Area-averaged

► Select Date Range: Set the date range from _____ to _____

► Select Region (Bounding Box or Shape): −180, −10, −80, 10. Click on the folded map icon to see the parameters (refer to Figure 2.4).

► In the Measurements drop-down menu, check the box for Chlorophyll, and then choose Chlorophyll α concentration (MODISA_L3m_CHL v2018) as the variable.

► Click on "Plot Data" (this may take a minute).

FIGURE 2.4

Image created using the Giovanni program. The highlighted area depicts the boundary region of the El Niño event.

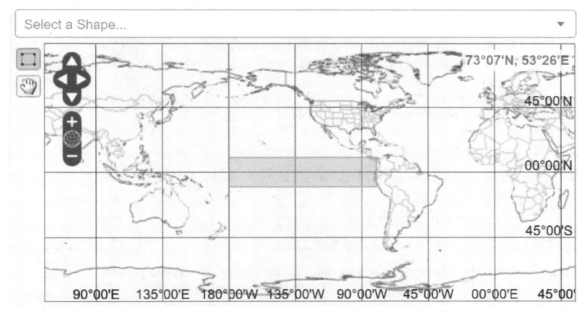

Source: Michelle L. Hardee. Background image is from NASA's Earth Observing System Data and Information System (EOSDIS)

Use Figure 2.4 and the plot you created to answer the following questions.

a. When did we begin to see the El Niño conditions occurring (month/year)? _____

b. How low did monthly chlorophyll α concentrations reach (with units)? When do you see these peak conditions occurring?

_____

_____

c. How long did the El Niño conditions last? _____

_____

6. Using the previous browser with the SVS website, search for the visualization #2078, **The Effect of the Amazon on the Atlantic** and play the animation.

a. Locate where the Amazon River empties into the western Atlantic. As the animation plays, watch the patterns of land biosphere changes, ocean color changes, and the primary productivity variation with time. What is the relationship between the river's drainage and seasons? _____

_____

b. What climatic factor(s), which are influencing the terrestrial biosphere in this region, do you think are contributing to the increase in ocean productivity at the Amazon River's mouth? _____

_____

c. How do these factors affect primary productivity in the ocean? _____

_____

7. Next, search for #2077, **SeaWiFS: The North Atlantic Bloom**. This link provides several years of ocean color data on the seasonal plankton bloom in the North Atlantic.

a. During what season(s) of the year (as viewed by the changes on land) does the plankton bloom in the North Atlantic cover the greatest ocean area? _____

_____

b. Why is such a large area of the North Atlantic Ocean so productive in that season?

_____

_____

c. In the Northern Hemisphere, what happens after the sea ice retreats every year? _____

_____

_____

8. Imagine you are a commercial fisherman that harvests tuna in the North Atlantic Ocean. How would you use satellite imagery to find the best fishing grounds? _____

_____

_____

9. If you were a local fisherman who occasionally went into inlet waters to catch nearshore fish, would you use satellite imagery as well? Why or why not? _____

_____

_____

10. Search for animation #30392, **Monthly Chlorophyll Concentrations**. Click on the animation from the Aqua/MODIS satellite from the past decade and answer the questions below. You can also use the link located under "For More Information" for extra help.

   a. Describe the new color scale. _____

   _____

   b. What is the average concentration of chlorophyll in coastal regions? _____

   What is the average concentration in the open ocean regions? _____

LAB 3

PRIMARY PRODUCTION

OBJECTIVES

- ▸ Learn how to accurately measure dissolved oxygen concentrations in aqueous solutions.
- ▸ Understand the relationships between dissolved oxygen, carbon, and primary production.

POINTS TO PONDER

- ▸ Why are oxygen concentrations higher than expected in oceanic surface waters?
- ▸ Why do hypoxic (low O_2) or anoxic (no O_2) zones develop in the coastal ocean?

TERMS TO KNOW

anoxic	net primary production	phytoplankton
dissolved oxygen	oligotrophic	primary producer
eutrophic	oxygen minimum layer	primary production
gross primary production	photoautotroph	primary productivity
hydrography	photosynthesis	Q_{10} effect
hypoxic	photosynthetic quotient	respiration

THE RELATIONSHIP BETWEEN OXYGEN AND PRIMARY PRODUCTION

Even though a great deal of energy (in the form of sunlight of various wavelengths) strikes the earth, only around 1% of that light energy reaches living systems. For this energy to be made usable, it must first be captured by **photoautotrophs**, organisms capable of making their own food through photosynthesis. **Photosynthesis** is a complex series of chemical reactions that can be summarized as follows:

$$6CO_2 \;+\; 12H_2\overset{*}{O} \xrightarrow{\text{Solar energy}} (CH_2O)_6 \;+\; 6H_2O \;+\; 6\overset{*}{O}_2$$
$$\text{(Carbon dioxide)} \hspace{4cm} \text{(organic material)}$$

This process results in the synthesis of glucose from water and carbon dioxide, with the *release* of dissolved oxygen as a by-product. As a result, organic products result from inorganic reactants. The predominant organisms that perform photosynthesis in the ocean are the plants and the single-celled algae and cyanobacteria called **phytoplankton**. They are called **primary producers** because they directly absorb solar radiation and use it to drive chemical reactions that result in "fixed carbon"— six-carbon sugars that form the building blocks of all life in the ocean. The total amount of organic material produced in the sea by photosynthesis (assuming that chemosynthesis is negligible) is termed **gross primary production** or GPP.

However, because photosynthesis is an energy requiring process, the solar energy converted must be partitioned, with part going to run more photosynthesis, and part going to storage (from which the rest of the living world can then obtain this stored energy), or to other metabolic processes in the plant, such as respiration. **Respiration** is a complex metabolic process that can be described by the following formula:

$$(CH_2O)_6 \;+\; 6H_2O \;+\; 6O_2 \longrightarrow 6CO_2 \;+\; 12H_2O$$

This equation should look familiar to you—it is basically the reverse of photosynthesis. All organisms, whether they are an autotroph or a heterotroph, perform respiration, in which the organic material is broken down by oxygen, releasing CO_2, water, and heat energy as by-products. Thus, the amount of organic material remaining after removing and accounting for that used for respiration is referred to as **net primary production** or NPP. This represents the amount of organic material *available to support other organisms.*

Other terminology is used to quantify the photosynthesis from primary producers. The amount of phytoplankton tissue that accumulates from photosynthesis over time is termed **primary production**, and the *rate* of phytoplankton growth and biomass increase is called **primary productivity** (in g $C/m^2/d$ or other similar units). This production rate determines the amount of food available for nearly all other organisms in the sea. The majority of primary production in the oceans is accomplished by the small, open ocean phytoplankton rather than large marine plants such as seaweed or salt marsh grasses—these only account for 5 to 10% of total marine productivity. A similarly small fraction of primary production comes from chemosynthetic organisms (using chemical energy instead of light energy); we will ignore these for now.

During the day, when light availability is highest and therefore photosynthesis is at a maximum, dissolved oxygen accumulates in the surface ocean. The **dissolved oxygen** (written as dO_2) is that amount of oxygen that is actually *dissolved* into the water (not existing as bubbles) and is available for use by organisms. The processes of photosynthesis and respiration by plants and animals are important factors causing fluctuations in the amount of dissolved oxygen present in the water. In waters having abundant growth of phytoplankton, wide diurnal fluctuations can be seen in the dO_2 concentrations. In particular, during daylight hours, the water in such areas may become supersaturated with oxygen. At night when organisms respire, dO_2 is removed from ocean waters and concentrations can dip to very low levels. Sustained dO_2 concentrations below 2 mg/L are problematic to most fish and some invertebrates. In coastal oceans, **hypoxic** (< 2 mg/L) or **anoxic** (no oxygen) bottom water conditions may develop that result in the death of marine organisms like fish and shellfish. The Gulf of Mexico near the outflow of the Mississippi River is famous for its "Dead Zone" which is caused by these kinds of conditions. In the **oxygen minimum layer** of the deep ocean (at around 800 meters), the concentration may be reduced to 50% of saturation. This depth receives essentially no sunlight, so oxygen is not added to the water through photosynthesis. However, microbial organisms and other marine animals live in this layer to feed on the abundant plankton material falling out of the surface waters. As these animals respire, they use up the oxygen, but there are no phytoplankton to renew the oxygen supply in this region, resulting in an oxygen minimum. The only method of renewal comes from vertical mixing.

FACTORS AFFECTING PRIMARY PRODUCTION

As you have already learned, there are numerous factors that influence the rate of primary production occurring in the oceans. Primary productivity is typically expressed as the total quantity of carbon fixed by phytoplankton, and therefore measured in grams of carbon fixed per square meter of sea surface over a unit of time ($gC/m^2/d$). Factors affecting this carbon fixation on a physical, ecological, or physiological level will limit or enhance primary productivity.

- ▶ *Light:* Photosynthesis is only possible when light reaching phytoplankton is above a certain intensity. Light is influenced by a number of factors such as clouds and dust. It is also influenced by the water itself due to reflection, scattering by suspended water particles (turbidity), and absorption of various wavelengths. The amount of light also varies with latitude, decreasing from the equator toward the poles as a result of the changing angle of Earth's surface with respect to the incoming solar rays.

▶ *Hydrographic conditions:* The **hydrography** of a region includes physical factors that move water masses around in the oceans, such as winds, currents, and upwelling. These processes can push phytoplankton populations below a critical light level required for photosynthesis, and reduce the amount of primary production occurring. Other parameters such as salinity, temperature, and oxygen levels can alter the growth rate of the photosynthetic cells and reduce productivity.

▶ *Temperature:* The temperature of the water column in which phytoplankton reside can affect both the physical aspects of the ocean environment as well as the physiology of the phytoplankton. Temperature stabilizes the water column during summer, preventing phytoplankton from being mixed to greater depths where there is less light. Temperature also directly influences the metabolism of the phytoplankton cell: metabolic rates double with a 10°C increase in temperature, called the "**Q_{10} effect.**" The Q_{10} temperature value is a measure of the rate of change of a biological system as a consequence of increasing the temperature by 10°C. It is a unitless quantity, as it is the factor by which a rate changes.

▶ *Nutrients:* The major inorganic nutrients required by phytoplankton are nitrate, phosphate, and silica. These occur in small amounts (mmol to µmol/L concentrations or less) and are important limiting factors for primary productivity. Trace metals, such as iron and zinc, are also important. **Oligotrophic** regions (such as the open ocean) have low concentrations of essential nutrients and therefore low productivity, while **eutrophic** waters (such as coastal areas) contain high nutrients and support high numbers of phytoplankton.

▶ *Grazing:* Many zooplankton feed upon the phytoplankton—these are the herbivorous zooplankton. In many oceans, zooplankton can graze phytoplankton cells as quickly as they accumulate, such that phytoplankton "blooms" are rarely observed.

PURPOSE OF THE LAB EXERCISE

A major area of research in marine community ecology is the determination of how much energy is captured by the primary producers in an area, and how external forces such as temperature, light, nutrients, etc., influence this rate of energy capture. This exercise will teach you a method for determining the amount of energy being fixed in a planktonic community, and the effect of temperature on the rate of photosynthesis.

As you can see from the photosynthesis equation above, oxygen is released as a byproduct. By measuring the amount of oxygen evolved, one can obtain an *indirect estimate of the amount of carbon fixed in photosynthesis* over a given period of time. While this seems straightforward enough, because plants and algae are living organisms, they also use up this energy via respiration, which requires the uptake of oxygen. In order to make an accurate estimate of the amount of carbon fixed, you must account for the oxygen evolved during photosynthesis that was then used up during respiration. Thus, a generalized equation would look something like this:

Total Carbon Fixed   =   Carbon into storage and growth   +   Maintenance
(Gross Primary Production)   (Net Primary Production)   (Respiration)

From an experimental viewpoint this can be rewritten as:

$$\text{Total } O_2 \text{ evolved} = [(O_2 \text{ measured in system at Time 1}) - (O_2 \text{ at Time 0})] + O_2 \text{ consumed}$$

One of the simplest methods of determining primary production is the light and dark bottle method. This method makes use of the changes in oxygen concentrations that occur while the organisms are confined in closed containers. In the dark bottle, no photosynthesis is occurring, so changes in oxygen in the bottle are due only to respiration. In the light bottle, both photosynthesis and respiration are occurring. If one knows the concentration of oxygen in the bottles at the beginning of the experiment ("Time 0"), and measure oxygen concentration in the bottles at some later time ("Time 1" or "Final"), it is then possible to calculate both the gross and net primary productivity. The details of this are covered in Part 2.

HOW AN OXYGEN METER WORKS

A common procedure used to determine oxygen content of water is the Winkler titration. This is an involved, but highly accurate procedure. In this laboratory, you will use dO_2 meters, which are simpler to use (but not as accurate). Standard oxygen probes use membrane-covered polarographic sensors (measurements using polarized or electrochemical differences) with built-in thermistors which measure and compensate for temperature. A thin, permeable membrane stretched over the sensor isolates the sensor elements from the environment, but allows oxygen and certain other gases to enter. When a polarizing voltage is applied across the sensor, oxygen that has passed through the membrane reacts at the cathode (where electrons are generated), causing current to flow. The membrane passes oxygen at a rate proportional to the pressure difference across it. Since oxygen is rapidly consumed at the cathode, the oxygen partial pressure inside the membrane can be assumed to be zero. Thus, the force causing the oxygen to diffuse through the membrane is proportional to the absolute pressure of oxygen outside the membrane. If the oxygen partial pressure increases, more oxygen diffuses through the membrane and more current flows through the sensor. A lower partial pressure results in less current. The meter is calibrated by measuring oxygen in fully saturated and completely deoxygenated fluids.

EXERCISE 3.1
dO$_2$ MEASUREMENTS USING LIGHT AND DARK BOTTLES

Materials required: Approximately 1 L of phytoplankton sample per section, 5 dO$_2$ meters with membrane-sensor probes, ~15 L of filtered seawater, 2 fluorescent light ballasts with 40-watt bulbs, aluminum foil and window screening (to adjust light intensity), 9 clear Biological Oxygen Demand (BOD) bottles with stoppers, filtering apparatus, GF/C filters, 1-mL pipette, paper towels

OBJECTIVE

The goal of this exercise is to measure and compare the primary production of one species of phytoplankton (your instructor will tell you the species) under two light treatments by measuring the amount of dissolved oxygen (dO$_2$) evolved by the cells over time.

DEFINE YOUR HYPOTHESIS

What do you expect to see concerning the *rate* of oxygen produced between the two different light treatments?

_____

_____

Knowing the information above, now define a hypothesis:

I hypothesize that _____

_____

A. METHODS

Experimental Setup

In a set of clear bottles, a series of "treatments" has been prepared to examine the effect of different light intensities (high vs. low light) on the oxygen production by the phytoplankton and, consequently, the rate of primary productivity. Seawater was filtered prior to adding the algal cultures to remove any larger organisms or material which might affect oxygen values. Cultures of the algae were carefully decanted into the bottles, to ensure a minimum amount of aeration.

Oxygen concentrations in the bottles prior to their incubation were determined. These are the "Time 0" values. Three bottles (replicates) were left uncovered and exposed to a direct light source ("High Light" treatment), three were left uncovered and shaded by a screen ("Low Light" treatment), and three were covered in aluminum foil to prevent light entering the bottle ("Dark" treatment). Each bottle has been labeled with its respective light intensity and replicate number (1, 2, or 3), and each bottle is filled to a zero headspace capacity. **In order to give the cultures time to carry out sufficient photosynthesis for oxygen to accumulate in the bottles, the cultures in your experiment have been incubating for several hours prior to lab.** *Be sure to find out how long the bottles have been incubating!*

At the end of the incubation period, you will measure oxygen concentration in all bottles ("Final measurement," which will be a number of hours depending on when your experiment incubation started), as well as chlorophyll *a* from each replicate of all treatments to be analyzed in a later lab.

PART 1: MEASURING OXYGEN

Ideally, every sealed treatment bottle would have an oxygen probe inserted into it. Because this is not the case, you must take special care to avoid introducing air into the bottles as you move the oxygen probe from bottle to bottle.

1. Carefully measure the oxygen content of all bottles—this is your Final Oxygen concentration measurement. Note the time for each measurement and record this information in Table 3.1.

2. Obtain the following information from your Instructor and complete Table 3.1:

   a. The starting times of the incubation, when the algae were *first* placed in the experiment bottles.

   b. The initial dO_2 readings of each treatment and replicate already recorded for you.

3. Take a 1-mL subsample from each replicate and filter for chlorophyll *a*. Place each filter into a clearly labeled vial with: section number, treatment, and replicate number, and place it in the freezer for analysis next week.

4. Carefully combine all treatments and replicates by gently pouring each bottle together into one container. Next, add 500 mL of filtered seawater to the container. Then refill the replicates for each treatment, filling the bottles up to the mid-neck line (so there is no headspace).

5. Carefully measure the oxygen content of all bottles—this is the Initial Oxygen concentration measurement for tomorrow's lab section. Note the time for each measurement and record this information in Table 3.2.

TABLE 3.1

Data from oxygen measurements for algal primary productivity.

Bottle #	Light Intensity	Time of Initial O_2 reading	Initial O_2 reading at Time 0 (mg/L)	Time of Final O_2 reading	Final O_2 reading (mg/L)	Time elapsed (hours)
1	High					
2	High					
3	High					
1	Low					
2	Low					
3	Low					
1	Dark					
2	Dark					
3	Dark					

TABLE 3.2

Initial oxygen measurements for algal primary productivity for the next lab section.

Bottle #	Light Intensity	Time of Initial O_2 reading	Initial O_2 reading at Time 0 (mg/L)
1			
2			
3			

PART 2: CALCULATIONS AND DATA ANALYSES
A. Calculating Rates of Primary Productivity

1. **NPP** values are determined directly from the "light bottles" incubated at the two light intensities and the "dark bottles". Do this in the space provided below using the equations as follows:

    $NPP_{low\ light} = ([Final\ O_2]_{low} - [Initial\ O_2]_{low}) \div time\ (h)$

    1.

    2.

    3.

    Average =

    $NPP_{high\ light} = ([Final\ O_2]_{high} - [Initial\ O_2]_{high}) \div time\ (h)$

    1.

    2.

    3.

    Average =

    $NPP_{dark} = ([Final\ O_2]_{dark} - [Initial\ O_2]_{dark}) \div time\ (h)$

    1.

    2.

    3.

    Average =

2. Calculate **GPP** by using the average NPP values calculated above in the following equations:

    Avg. $NPP_{low\ light}$ + [Avg. Respiration (NPP_{dark})] = GPP_{low}

    Avg. $NPP_{high\ light}$ + [Avg. Respiration (NPP_{dark})] = GPP_{high}

B. Convert Using a Photosynthetic Quotient to Quantify Photosynthesis

You will now *quantify* the amount of photosynthesis that occurred in the light treatments by calculating the amount of carbon "fixed" (CO_2 converted into organic carbon) from the quantity of dissolved oxygen evolved (produced by photosynthesis).

1. Using your GPP value calculated above for each light intensity (the average for your lab section), convert this oxygen value into the relative amount of carbon fixed, which is a more common way of evaluating the rate of productivity. To do this, first convert your rate of O_2 produced (mg O_2/L/h) into units of "mmoles O_2/L/h" by dividing by the molecular weight of O_2 (32 g/mole). Do this in the space below.

   Low Light treatment:

   High Light treatment:

2. Next, divide this value by the **Photosynthetic Quotient (PQ)**. The PQ is simply a conversion factor that gives **the moles of oxygen produced, divided by the moles of carbon dioxide assimilated.** (Or, in our case mmoles of oxygen produced divided by the mmoles of carbon dioxide assimilated.) PQ values typically range from 1–2, and vary from 1.0–1.3 for marine single-celled algae. *Use an average value of 1.2 in your calculation.* Show your work below—make sure you end up with the proper units!

   Low Light treatment:

   High Light treatment:

3. Now convert the mmol CO_2/L/h to mg C/m^3/h:

Low Light treatment:

High Light treatment:

4. What do these calculated values (mg C/m^3/h) represent. _____

_____

B. RESULTS

1. Based on your rate calculations, what is the *average* rate of O_2 production for each of the treatments for your lab section, from the time the algae were originally placed in the bottle until your final dO_2 measurement?

    Low Light: _____ High Light: _____ Dark: _____

2. What was the total primary productivity for each of your treatments?

    Low Light: _____ in units of _____

    High Light: _____ in units of _____

C. DISCUSSION

1. What is the difference between phytoplankton biomass and phytoplankton primary productivity?

_____

_____

_____

2. Revisit your hypothesis at the beginning of this exercise. Did you observe what you expected for the different treatments? Why or why not? Explain.

_____

_____

_____

3. One basic assumption is made for this experiment. What is it? (Hint: It has to do with respiration of the available oxygen in the bottles.)

_____

_____

4. What are some other factors that might have influenced your oxygen values in the two light treatments?

_____

_____

5. What is the physiological explanation for why the rate of photosynthesis (and therefore oxygen evolved) changes as the level of light available for photosynthesis is varied? (You may need to look this up in your textbook or other source for the answer.)

_____

_____

6. Describe a procedure you would use to determine primary production by phytoplankton at four depths from surface to bottom in the mouth of the North Inlet estuary along the coast of South Carolina.

_____

_____

_____

_____

_____

_____

LAB 4

EXPERIMENTAL AND STATISTICAL ANALYSES

OBJECTIVES

► Use graphs and statistics to make comparisons between different communities of marine organisms.
► Develop skills in organizing and presenting data, and analyzing results.

POINTS TO PONDER

► How do scientists make sense of the data they collect?
► On what type of information do marine conservationists base their management decisions?
► How much data is "enough"?

TERMS TO KNOW

confidence interval	normal distribution	sample size
confidence level	null hypothesis	sampling
confidence limits	paired data	standard deviation
estimate	parameter	standard error
hypothesis	percentiles	statistic
linear regression	p-value	statistical analysis
mean	random	*t*-test
median	range	variable
mode	R-squared	variance

THE PURPOSE OF EXPERIMENTAL AND STATISTICAL ANALYSES

In the sciences, research often focuses on experiments in which a variable is altered or controlled by the investigator. A **variable** is a factor or condition that can be controlled or changed in some measurable way. For example, say we measured the individual shell color and length of a set of marine snails. Color and length are two separate properties, or variables, each with their own range of variation. Experiments are designed based on the hypothesis or hypotheses being investigated. Experimental analysis is the concept of testing the probability that a hypothesis is correct. To do this, we incorporate a **statistical analysis**, or statistics, into our research study. **Statistics** is the scientific *study of data*—the measurement of naturally varying factors in order to quantify the error within that natural variation and predict it in future measurements. Statistics allows us to determine whether the regular occurrences of small differences measured in the data are real and valid. Further, it allows sub-sampling of large groups of individuals without having to separately count and analyze every single individual. Many areas of research in addition to the sciences, from public opinion polls to quality control on manufactured items, rely on statistical analyses for determining the value of the data obtained. It is especially important that students of scientific disciplines become familiar with statistical methods as early as possible. This laboratory exercise provides you the opportunity to actively learn some of the basic methods used in statistical analysis using data you have produced and recorded. Today you will perform two different statistical analyses: a test of difference and a test of relationship.

DATA COLLECTION

The process of data acquisition is called **sampling**. There are many ways to sample; however, the rules for sampling remain consistent. One obvious rule is that the more observations you acquire, the better your prediction of a value. When comparing groups, balanced sampling is important (i.e., collecting the same number of observations from each group). Also key is ensuring that your data collection is as **random** as possible. This sounds simple, but acquiring a random set of samples is not always easy. An unbiased estimate of a population is only possible if the total population is truly represented in the sample units. There are a wide variety of sampling strategies available and several steps involved in the sampling strategy. Developing a strategy includes: choosing a sampling unit (e.g., a quadrat, transect, or individuals); determining a **sample size**, "n", the number of observations or replicates to include in a statistical sample; arrangement of random sampling (e.g., divide an area into equal plots and sample equally at random); and timing of sampling (if important to the hypothesis, e.g., tidal, seasonal, diel).

THE NORMAL DISTRIBUTION AND THE MEAN

As you know, in nature there is a great deal of variability in almost all characteristics of organisms. If you were to measure the height of students in your class, you would expect to find a few very tall people and a few very short people, while the majority of the class would fall somewhere between these extremes. This type of distribution is known as a **normal distribution** and is one of the most common types of distributions found in nature. If this distribution were plotted on graph paper, it would look like a bell-shaped curve (Figure 4.1).

From the normal distribution, we can determine three basic parameters: the median, the mode, and the mean. When all the values are ordered sequentially, the **median** is the value located directly in the center, with equal numbers of larger and smaller values on each side. The **mode** is the most frequently occurring value, the value at which the distribution curve "peaks." The **mean** is the average of the population and is calculated by summing all the individual observations or items and dividing this sum by the number of items in the sample. Thus, if we denote the individual observations or items for a particular variable as "X", then the mean is calculated as:

$$Mean\ (\bar{X}) = \frac{\text{Sum of ``}X\text{''}}{n} = \frac{\Sigma X}{n}$$

The mean is symbolized by $\bar{X}$ (pronounced "x-bar"), the capital Greek symbol Σ (sigma) means to sum the items indicated (in this case all "X" values), and "**n**" is defined as the number of observations in a sample, or the **sample size**. Of the three basic parameters, the mean is typically the best estimate of the distribution of variability. It takes into account not only the most frequently occurring heights (using our previous example), but the number of times each of the heights occurs as well as the extremes, thus giving a truer estimate of the typical height. In a perfect normal distribution, these three estimates are the same; however, "perfect" rarely, if ever, occurs in the real world.

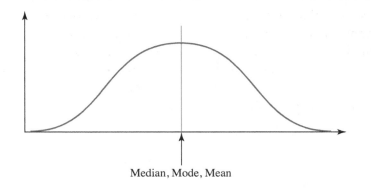

FIGURE 4.1

Typical bell-shaped curve for a characteristic that shows a normal distribution. In this ideal diagram, the median, mode, and mean (see text) all lie at the same point, though this is not necessarily true for all normal distributions.

Median, Mode, Mean

Source: Michelle L. Hardee

VARIANCE AND STANDARD DEVIATION

Referring back to our height example and the normal distribution in Figure 4.1, you know that not all the heights have the actual average value, so the distribution is composed of a range of values on either side of the mean. This brings us to the next important statistical measure, the **standard deviation (SD)** of a value from the mean. The standard deviation measures how accurate the mean is as an estimate of a typical value. It does this by measuring the distribution of all observations from the mean and weighting each item by its distance from the center of the distribution. It is essentially an "average" deviation of a given value away from the mean. This statistic is valuable because a mean can sometimes be misleading as an estimate of a typical value. Examine the following example:

	Example #1:	90	Example #2:	125
		80		100
		70		75
		+ 60		50
		300		+ 25
				375
Mean ($\overline{X}$):		300 ÷ 4 = 75		375 ÷ 5 = 75
SD:		= 12.9		= 39.5

While both means are the same, the distributions of individual values are not. In one case, the spread or **range** of values is relatively small, while in the other the range is much greater and has a larger sample size. Thus, in Example #1, the mean of 75 is a truer **estimate** of the typical value than in Example #2. This is easy to see in this example, but if you were only presented with the means, you would not see a difference between the two distributions. The usefulness of the standard deviation is that it gives you a quantitative idea of how accurate an individual value is relative to the mean.

The standard deviation is calculated through a series of steps. First, calculate the deviation "d" by subtracting the mean from each sample value ($X - \overline{X}$), then square each of these deviations and sum all of the squared deviations (Σd^2). Next, divide this number by the sample size minus 1 (denoted as "n – 1"). This value, the average of the squared deviations, is called the **variance** and is a measure of fundamental importance used in numerous types of statistical analyses. The standard deviation is then obtained by taking the square root of the variance. Study the examples below so you can understand how variance and standard deviation are calculated.

Example #1:   **Example #2:**

X	$(X - \bar{X}) = d$	d^2	X	$(X - \bar{X}) = d$	d^2
90	15	225	125	50	2500
80	5	25	100	25	625
70	−5	25	75	0	0
60	−15	225	50	−25	625
			25	−50	2500
$\bar{X} = 75$		$\Sigma d^2 = 500$	$\bar{X} = 75$		$\Sigma d^2 = 6250$

n = 4, so **(n − 1)** = 3   **n** = 5, so **(n − 1)** = 4

variance = $\dfrac{\Sigma d^2}{n-1} = \dfrac{500}{3} = 166.7$ **variance** = $\dfrac{\Sigma d^2}{n-1} = \dfrac{6250}{4} = 1562.5$

SD = $\sqrt{\text{variance}} = \sqrt{166.7} = 12.91$ **SD** = $\sqrt{\text{variance}} = \sqrt{1562.5} = 39.53$

In literature you will see it written as: Example #1 had an average of 75 ± 12 (s.d.) and Example #2 had an average of 75 ± 39 (s.d.). This tells the reader that there was more variability in Example #2.

As you can see, the standard deviation is an approximation of the average difference (deviation) of the values from the mean. The smaller the standard deviation, the closer the sample values are to the mean, and in turn the more accurate the sample mean is as a population estimate. When sampling in nature, it would be useful to distinguish between the sample mean that we calculated from, for example, a small sample of phytoplankton abundances collected off Folly Beach pier, and an entire population of phytoplankton abundance along the southeastern U.S. shoreline. However, in reality we cannot realistically measure either directly. A value *computed* entirely from a *sample* is called a **statistic**, while a *measurable characteristic* of the *total population* is called a **parameter**. The true value of most population parameters is typically unknowable; this is why we must use sample statistics to estimate them.

STANDARD ERROR AND CONFIDENCE INTERVALS

The last two important statistics computed in this exercise are the standard error and confidence intervals. The **standard error** (**SE**) is typically calculated on sample statistics such as means instead of individual observations. It gives us the *standard deviation of a distribution of means* calculated from samples of a given sample size, **n**. Essentially, it is the expected deviation of other means if the variable was sampled many times, all producing slightly different means (usually this is not done, as standard error is calculated to provide that information). Think of it as the standard deviation of several means determined from the original sample set. The standard error is calculated by taking the square root of the variance divided by the number of values (**n**):

$$\mathbf{SE} = \sqrt{\frac{\text{variance}}{\text{n}}}$$

Using the examples given previously:

Example #1:

$$SE = \sqrt{\frac{166.7}{4}} = 6.5$$

Example #2:

$$SE = \sqrt{\frac{1562.5}{5}} = 17.7$$

An important attribute of both the standard deviation and standard error in a normal distribution is that 68.3% of the sample values fall within plus or minus (±) one standard deviation of the mean (± 1 SD), or for means, one standard error of the sample mean (± 1 SE). This corresponds to 34.13% of the area on either side of the mean under the normal distribution curve. Such percentages are called **percentiles**, and represent the area under a normal distribution where such a sample statistic may be contained. Examine the percentiles as shown in Figure 4.2.

From Figure 4.2,

$\bar{X}$ ± 1 SD contains 68.27% of the items
$\bar{X}$ ± 2 SD contains 95.45% of the items
$\bar{X}$ ± 3 SD contains 99.73% of the items

FIGURE 4.2

Parameters for the population, the mean, and standard deviation (**s**, defined as SD in this lab), and their corresponding areas under the normal distribution curve.

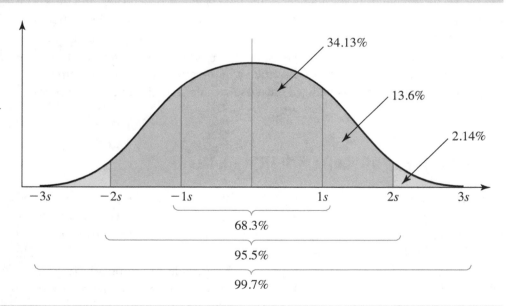

And conversely,

> 50% of the items fall between the mean ± 0.674 SE
> 95% of the items fall between the mean ± 1.960 SE
> 99% of the items fall between the mean ± 2.576 SE

Why are these areas under the normal curve important? As mentioned in the previous section, the true values of the population parameters almost always remain unknown, because it is simply impossible to sample the entire population. So we estimate the reliability of a sample statistic such as a mean (i.e., how close it is to the population mean) by providing a range limit to it. It allows us to answer the question, "Within what range can we be reasonably sure that the population mean resides?" The standard error (or standard deviation) can provide this range of values for this variable of interest. Such a range of values is called the **confidence interval**, and can tell us how likely it is to capture the true—but unknown—size of the observed variable. This range also has a specified numerical probability of its likelihood of including the true value of the variable. The specified probability is called the **confidence level**, and the end points of the confidence interval are called the **confidence limits**. By convention, we create confidence intervals at the 95% level—this means that 95% of the time a confidence interval should contain the true value of the variable of interest. In other words, the confidence interval provides a range for our estimate of the true value and gives us an idea of whether the population mean is close to our sample mean.

Percentiles and confidence intervals can be applied to both standard deviations and standard errors, but by using the standard error of the mean, it is possible to determine with a high degree of probability the value of the true population mean. For example, if you take the sample mean and either add or subtract one standard error, there is a 68.27% probability that the population mean will lie somewhere within this range of values. If we use the value mean ± 0.674 SE, we know that 50% of the time our population mean is enclosed by those limits. With a value of mean ± 3 SE, we know that there is a 99.73% chance that our population mean is included within these limits. The following shows the calculation of these percentiles using the mean and standard error from the previous examples:

Example #1:

68.3%: $\bar{X} \pm 1$ SE = 75 ± 6.5
            = 75 − 6.5 & 75 + 6.5
            = 68.5 to 81.5

95%: $\bar{X} \pm 1.960$ SE
         = 75 ± (1.960 × 6.5)
         = 75 − 12.74 & 75 + 12.74
         = 62.3 to 87.7

Example #2:

68.3%: $\bar{X} \pm 1$ SE = 75 ± 17.7
            = 75 − 17.7 & 75 + 17.7
            = 57.3 to 92.7

95%: $\bar{X} \pm 1.960$ SE
         = 75 ± (1.960 × 17.7)
         = 75 − 34.69 & 75 + 34.69
         = 40.3 to 109.7

Thus for Example #1, there is a 68.3% probability that the true population mean will fall between 68.5 and 81.5 and a 95% probability it will fall somewhere between 62.3 and 87.7. On the other hand, in Example #2 the 68.3% probability confidence interval is between 57.3 and 92.7, and the 95% range is between 40.3 and 109.7. As you can see, with a 95% probability range of {40.3 to 109.7} points, any estimates of the population mean based on the sample in Example #2 would not be much better than a guess. From this example, you should be able to see that the standard error is a helpful statistic to judge how good an estimator your sample is of the true population.

HYPOTHESIS TESTING

A majority of scientific investigations are based on the idea of hypothesis testing. The idea is that you formalize a **hypothesis** (H_1), a proposed explanation for a phenomenon, into a statement that can be determined true or not by experimental testing. Every hypothesis will have an associated **null hypothesis** (H_0, read "H-naught") that most statistical tests use. The null takes the "default" side that there is no difference or no relationship between two measured phenomena.

A statistical test uses a **p-value** to determine the probability that the null hypothesis is true. If the probability is low then the null hypothesis is rejected and the original hypothesis is accepted. Generally, a 95% probability is used to determine if a hypothesis should be rejected or accepted, however a 99% probability can be used to determine a more confident result. When using a 95% probability, the p-value that is calculated will be considered either greater than (>) or less than (<) 0.05 (0.01 for 99% probability). If the p-value is > 0.05 the null hypothesis should be *accepted*; a p-value of < 0.05 indicates you should *reject* the null hypothesis.

There are many different statistical tests to determine whether observations from two groups differ significantly (rejecting the null hypothesis). The choice of test depends on the type of data you collected. The data we will be using today is considered **paired data** or matched data. Paired samples occur when a single individual is tested twice (e.g., repeated, or before and after) or the sample site is tested more than once. Paired data also occurs when individuals of a clone are divided and subjected to two treatments, or a comparison of two population means. Today, we will be using the *t*-**test** to test our hypothesis. This test is used when the variances of two normal distributions are unknown, and when an experiment uses a small sample size.

LINEAR REGRESSION ANALYSIS

Linear regression is an extremely powerful and useful technique that is widely used in statistics. It determines the form and strength of a relationship between two variables, in some cases one observation being the 'cause' (independent variable) and the other being the 'effect' (dependent variable). It can describe the percentage of the prediction in the 'effect' variable, (i.e., how well can you predict a y-value if you only have an x-value) or the percentage describing the amount of variation in the 'effect' accounted for by the 'cause' (i.e., how certain you are that the y-variable is caused by the x-variable). In other cases it describes a simple correlation between two variables (with no causation assumed).

When dependent and independent variables are plotted, the 'slope' of the data distribution is the slope of a straight line of best fit drawn through the set of points defined by two variables. The slope can be positive or negative indicating an increase or decrease of y with an increasing value of x. A slope of zero indicates no change in y with x, and therefore no relationship. Once a best fit line is determined, then a value for the 'effect' can be predicted for any value of the 'cause'. An R^2 (pronounced "**R-squared**") value is a common output either expressed as a number between 0 and 1 or as a percentage. It is a statistical measure of how well a best fit line (regression line) approximates the distribution of real data points. An R^2 of 1.0 (100%) indicates a perfect fit. A p-value is calculated to determine if the slope is significantly different from zero ($p < 0.05$), indicating a significant relationship between the two variables.

STATISTICAL SYMBOLS AND TERMS TO KNOW

n = number of individual counts in a sample, or the sample size

Σ = to sum a series of values

$\bar{X}$ = the average (mean) of the samples

SD = the standard deviation of the mean

SE = the standard error

p-value = determines the significance of the test
         < 0.05 – There is a significant difference
         > 0.05 – There is no significant difference

R^2 = determines the relationship between two variables
         < 0.5 – There is a poor relationship: less than 50% of variable y can be explained by variable x.
         > 0.5 – There is a good relationship: more than 50% of variable y can be explained by variable x.

AT-HOME PORTION INSTRUCTIONS

Appendix B in this lab manual provides a brief introduction on how to use Microsoft Excel if you have never used the program. **Before you come to lab**, read and follow along with the procedures to become familiar with the Excel program. <u>**You will use what you learn in that Appendix in lab this week**</u>.

EXERCISE 4.1
PRIMARY PRODUCTIVITY DATA ANALYSIS

Materials required: Excel spreadsheet from Lab 3, 90% acetone, fluorometer, glass test tubes

DEFINE YOUR HYPOTHESES

In this exercise you will be asking two questions:

1. Is there a significant difference in the rate of oxygen produced between the two different light treatments from Lab 3?

2. Is there a significant relationship between the concentration of chlorophyll *a* in the phytoplankton cultures used for Lab 3, and the corresponding rates of oxygen production for these samples?

What will be the two null hypotheses for each question you are testing?

1. H_0: _____

   _____

2. H_0: _____

   _____

A. METHODS

1. Locate the vials for chlorophyll α analysis that you filtered last week. Your instructor has already dispensed 5 mL of 90% acetone into each vial and left to extract in the freezer overnight. Fill in the correct volumes into the spaces provided in Table 4.1.

2. Allow the vials to warm to room temperature and measure each replicate using the calibrated fluorometers. Report the values in Table 4.1.

TABLE 4.1

Information needed to calculate chlorophyll a concentrations.

	Replicate 1	Replicate 2	Replicate 3
Volume of seawater filtered (ml) [V_f]	1 mL	1 mL	1 mL
Volume of acetone for extraction (ml) [V_a]	5 mL	5 mL	5 mL
Fluorometer measurement (μg/L) [F]			

Use Microsoft Excel to execute the following analyses:

Null Hypothesis 1

3. Calculate the mean, standard deviation, and standard error of the oxygen production rates from each light treatment using the entire data set (see Appendix B for tutorial). Report these values in the results section (Question 1).

4. Using the data calculated in step 3 above, compare the average rates for oxygen production for each light treatment by plotting them on a bar graph. Include the error bars to show standard error. Label this graph as Figure 1. Answer Question 2.

5. Perform a *t*-test to test your Hypothesis 1 (see Appendix B for tutorial). Use the entire data set from each lab section and report the statistical analysis in the results section (Question 3).

Null Hypothesis 2

6. Using your specific lab section data, plot a linear regression to compare the rate of oxygen produced to chlorophyll a concentration of each light treatment (see Appendix B for tutorial). Label this graph as Figure 2. Answer Questions 4 and 5.

7. Continuing with the data used in step 6, perform a Model I linear regression analysis to test your Hypothesis 2 (see Appendix B for tutorial). Use the data from your specific lab section and report the statistics in the results section (Question 6).

B. RESULTS

Null Hypothesis 1

1. a. What are the mean and standard errors of oxygen production rates for the high and low light treatments? (e.g., 3.2 ± 0.2 with units)

   Low Light treatment:

   High Light treatment:

   b. Which light treatment had the highest variability?_____

2. Write a figure caption for your Figure 1 (the bar graph you made comparing the means for the two light treatments).

   _____

   _____

3. a. What is the calculated p-value from the statistical analysis you ran to test your null hypothesis 1?

   _____

   b. Did you accept or reject your null hypothesis 1?

   _____

Null Hypothesis 2

4. a. What is the slope and R^2 value calculated by the linear regression (Figure 2)?

   _____

   b. What percentage of oxygen production rate can be explained by the chlorophyll a concentration?

   _____

5. Write a figure caption for your Figure 2.

_____

_____

6. a. What was the calculated p-value from the statistical analysis you ran to test your null hypothesis 2?

_____

   b. Did you accept or reject your null hypothesis 2?

_____

C. DISCUSSION

Hypothesis 1

1. Did you expect there to be a significant difference in oxygen production rates between the two light treatments? Why or why not?

_____

_____

_____

Hypothesis 2

2. Explain why you would not expect to have a significant relationship between oxygen production rates and chlorophyll α concentration.

_____

_____

_____

3. What are some other experimental questions we could have asked and tested for during this experiment?

_____

_____

_____

LAB 5

THE "NUTS AND BOLTS" OF TAXONOMY

OBJECTIVES

- ▶ Understand how organisms are classified using the Linnaean System of taxonomy.
- ▶ Become adept at using a taxonomic key for identifying organisms.
- ▶ Learn to identify mollusks and fishes common to the South Carolina coast.

POINTS TO PONDER

- ▶ Why must every organism have a unique and specific Latin name?
- ▶ How are all species of organisms in the world categorized and classified?
- ▶ How can you use taxonomy to determine relationships and genetic ancestry in organism groups?

TERMS TO KNOW

apomorphic	homologous characters	phylum/phyla
biodiversity	ICZN	plesiomorphic
category	Linnaean System	polyphyletic
characters	meristic	species
classification	monophyletic	symplesiomorphy
dichotomous key	morphology	synapomorphy
extant	phylogenetic tree	taxon/taxa
genealogy	phylogenetics	taxonomic key
genus/genera	phylogeny	taxonomy

TAXONOMY AND THE LINNAEAN SYSTEM

The total number of species on Earth is estimated to range from 6–30 million, depending on how the estimates are made. The species alive today are only a tiny fraction of the billions of species that have existed on Earth since life first evolved, over 3.8 billion years ago. Further, over 75% of the described **extant** (living) species belong to the Phylum Arthropoda, which includes familiar organisms such as crabs, insects, spiders, and shrimp. Of these, the insects are by far the most abundant. So how do scientists categorize all these organisms? Is there any way to categorize them based on their evolutionary traits or relationships?

At first, attempting to classify organisms may seem to be a daunting task. The process of classifying organisms, or **taxonomy**, can be viewed as a puzzle, with straightforward rules to be followed. Most students think taxonomy is simply the process of scientifically naming organisms for the purpose of grouping and classifying them. In actuality, it is much more than that. The field of taxonomy includes the theory and practice of describing the diversity of organisms, and ordering this information into a logical system that reflects the *relationship* between the organisms. When grouping organisms, specific terms are used with which you should become familiar. A **category** designates a given rank or level in a hierarchical classification. Taxonomic levels such as species, genus, family, and order designate categories. A **taxon** (plural: **taxa**) is a grouping of organisms given a proper name, such as Phylum Arthropoda, or Class Mammalia.

The most difficult task of a taxonomist is assigning a taxon to the appropriate categorical rank, because the hierarchy of categories to which a taxon is assigned should reflect similarity of characteristics and descent from a common ancestor (see Figure 5.1). The most closely related organisms are grouped into the **species** category. A species is a group of potentially interbreeding natural populations that are repro-

FIGURE 5.1

A generalized arthropod phylogenetic tree. This tree represents a **monophyletic** scheme where a single ancestor is believed to have given rise to all classes of arthropods. Some phylogeneticists argue that arthropods should be in a **polyphyletic** tree in which the arthropod body type arose four separate times.

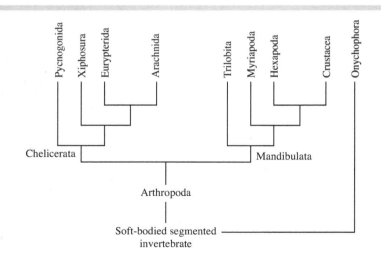

Source: Michelle L. Hardee

ductively isolated from other similar groups. The most closely related species are combined into a single **genus** (plural: **genera**), grouped together due to similarities in behavior, morphology, genetics, or other characteristics. Groups of related genera are combined into families, then into orders, classes, and **phyla** (singular: **phylum**). The category of phylum is distinguished by having a *unique body plan* and there are only about 32 animal phyla known to exist. To make matters more complicated, an organism is often further categorized by subfamilies, infraorders, suborders, super-orders, subphyla, etc.

Such a classification system based on hierarchical categories is known as the **Linnaean System** of Binomial Nomenclature, named after Carolus Linnaeus, the scientist who developed the method of using categories and a two-name system to name an organism. In addition to this system, a set of rules known as the International Code of Zoological Nomenclature, or **ICZN**, has been developed based on Linnaeus' system to provide universal guidelines for naming and classifying organisms. Here are a few of these fundamental rules which you should know:

1. When a new species is identified, it must be named in the binomial (two-name) form, and must be in Latin.
2. When writing the name of an organism, the genus and species name should always be italicized.
3. The generic (genus) name is always capitalized, but the species name is *never* capitalized.
4. The names of genera are required to be unique, but this is not true for higher taxa. As you can see in Table 5.1 below, the order name Decapoda occurs in both arthropods and mollusks. It is common to see identical species names as well.

As you can imagine, this classification system is not without problems, as taxonomists may categorize certain taxa differently. For example, some taxonomists place crabs in the Class Crustacea, while others place crabs in the Class Malacostraca of

TABLE 5.1

Examples of taxonomic classifications schemes for some common marine organisms.

	Eastern Oyster	Common Squid	Acorn Barnacle	Ghost Crab	Fiddler Crab
Phylum	Mollusca	Mollusca	Arthropoda	Arthropoda	Arthropoda
Class	Bivalvia	Cephalopoda	Maxillopoda	Malacostraca	Malacostraca
Order	Pterioida	Decapoda	Cirripedia	Decapoda	Decapoda
Family	Ostreidae	Loliginidae	Balanomorpha	Ocypodidae	Ocypodidae
Genus	*Crassostrea*	*Loligo*	*Balanus*	*Ocypode*	*Uca*
Species	*virginica*	*pealeii*	*tintinabulum*	*quadrata*	*pugilator*

the Subphylum Crustacea. Yet others view crustaceans as a Superclass. These classifications will no doubt continue to change as new information is obtained through detailed morphologic studies and genomic sequencing.

PHYLOGENETICS

A similar discipline, **phylogenetics**, differs from taxonomy in that it produces classifications that reflect the *genealogical relationships* between organisms. You may be familiar with your family **genealogy**, a "tree-like" representation of the historical lines of descent from one of your ancestors. A **phylogenetic tree**, or **phylogeny**, is similar to this, representing the historical course of speciation for a particular species or higher taxon (refer to Figure 5.1). A phylogenetic tree based on inherited character traits passed on to offspring creates a **classification** (including taxonomic categories) which represents the hypothesized evolutionary relationships among a group of organisms. This tree is a natural by-product of evolution, as it directly reflects inherited traits. The branching patterns in a phylogenetic tree are defined by the presence of unique, evolving innovations or characteristics (called *derived characters*, see below) shared by all members of that group (Figure 5.2). For the characters, phylogeneticists use a combination of fossil records, comparative anatomy and molecular data to understand the relationships among organisms and create the trees. However, often scientists do not know many of the tree branches, either the unique characteristic at the node or even other genetically related taxa. As scientists uncover new information, phylogenetic relationships between related organisms will continue to change.

FIGURE 5.2

A phylogenetic tree depicting one hypothesized view of the phylogeny of Phylum Mollusca. Dots represent synapomorphies that define each group. See text for explanation.

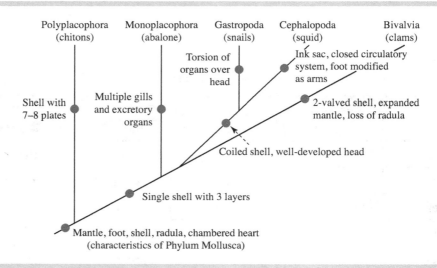

Source: Michelle L. Hardee

Organisms are placed in a phylogenetic tree based on similarities or differences in **characters**, a feature or observable part of an organism that can be described, repeatedly measured or counted, etc. There are many types of characters used in phylogenetic trees. Behavioral characters include features such as the mating dance of fiddler crab or burrow architecture of an intertidal worm. Genetic characters include the amino acid sequence of a particular protein, or the base pair sequence of the encoding gene (DNA). Morphological characters are anything that can be measured on the body of the organism. **Homologous characters** are features that have the same evolutionary origin, and therefore the same genetic basis. Homologous characters can be further defined by considering their presence or loss in species over time. The original, preexisting, or primitive character is called the **plesiomorphic** character, and the new, *derived* character that has arisen from it is called the **apomorphic** character. For example, the presence of a shell in all mollusks is a plesiomorphy, and its subsequent loss in slugs (a subclass within Class Gastropoda) is the new, derived apomorphic character. If a character found in two species is present in the *immediate* ancestor of these two species, this character is called a **synapomorphy** (Figure 5.2). If the character is present in the immediate ancestor but can be shown to have arisen in an earlier ancestor, this character is called a **symplesiomorphy**. The appearance of hair and mammary glands are synapomorphies unique to mammals, but are symplesiomorphies *within* the Class Mammalia. The identification of synapomorphies is the most important means for determining close evolutionary relationships between species. In order to determine the relationships between characters, use the following guidelines:

1. Identify the homologous characters among the organisms being studied.
2. Determine the direction of character change or "evolution" of the characters, particularly the synapomorphies.
3. Construct a phylogenetic tree of the taxa possessing all of the characters identified.

TAXONOMIC KEYS

The number of species of organisms in an ecosystem, or its **biodiversity**, is often used as a measure of an ecosystem's health. For example, the number of fish species on a coral reef or the number of polychaete worm species in marine sediments is a partial measure of biodiversity. In order to determine the number of species, you first must be able to distinguish one species from another. This is not an easy task! Species are genetic units, and individuals within a species can interbreed. However, it is impractical, if not impossible, to determine the *possibility* of interbreeding (particularly with fossils) of an individual to establish if it is the same as another. Thus, in practice, species are distinguished on the basis of different character **morphologies**, such as shell shape, size, number of appendages, etc. Taxonomists (and phylogeneticists) spend a significant part of their career attempting to determine which characters are "good"; that is, which characters will reliably allow them to distinguish one species from another. From this, taxonomists who have specialized in a particular group or species can develop a **taxonomic key**, or guide, to help others correctly identify individual organisms. There are four types of morphologic characters used in taxonomic keys:

1. **Presence/Absence:** A character is either present, or not. Two fiddler crabs of the genus *Uca* can be found in intertidal marshes. *Uca minax* is distinguished from *U. pugilator* by examining its enlarged pincer, which is colored white with red joints, whereas the pincer of *U. pugilator* is not.
2. **Color/Pattern:** Many characters deal with the external color of shells, skin, scales, or feathers. For example, the purse crab (*Persephona punctata*) is identified as having large, round, reddish-brown spots. No other crab has this particular color and pattern.
3. **Meristic:** Characters that can be counted and that will vary between species, such as the number of fin rays (fin bones) in fish dorsal fins, or numbers of legs or segments of crustaceans.
4. **Morphometric:** Characters of length. Because simple length measurements are typically a function of age—as organisms get older, they get larger—so morphometric measurements need to be expressed as a relationship or ratio between two measurements. For example: length of the head / length of the body.

The most common taxonomic keys are **dichotomous keys**, or two-forked, a style of taxonomic key that provides you with two choices at a time. They are constructed using contrasting pairs of statements that describe some feature or characteristic of the organism. For example, "*Exhibits pentamerous symmetry and tube feet*" or "*Lacks pentamerous symmetry and tube feet, possesses tentacles (with nematocysts)*." A decision must be made between each set of statements before proceeding. The process can be time-consuming and fraught with difficulty for the non-professional because uncommon terms are often used. However, if used properly, dichotomous keys lead to a single and correct identification, eliminating all other possibilities along the way. To use it correctly, you must read the possibilities in the very first category given and choose which of the two choices fits your organism best. A name or a number is given at the end of the description for your choice, which either identifies your organism or leads you to the next set of choices. This process is repeated until you obtain a name for your organism. Below is a simple example of a key using several sporting balls as the "species." The number in parentheses indicates the selection number from which you just came.

Key to several "species" of balls used in sporting events	
1 Ball round..	.2
Ball oblong and pointed at ends	Football
2(1) Ball solid..	.3
Ball hollow..	.4
3(2) Ball with 3 holes, diameter >15 cm, made of plastic-like material................	Bowling
Ball without holes, diameter < 10 cm, surface of cowhide	Baseball
4(2) Ball made of rubber, > 4 cm diameter	.5
Ball made of plastic, < 4 cm diameter	Ping-Pong
5(4) Ball with fuzzy surface..	Tennis
Ball with rubberized surface	Basketball

Dichotomous keys are quite useful tools for helping to identify different organisms, and are usually found in field guides, because identification in the field is based on features that are observable to the eye. It is important to remember, however, that a taxonomic key is an identification tool only, and is not synonymous with phylogenetic trees, which represent hypothesized evolutionary history.

EXERCISE 5.1
DICHOTOMOUS KEYS

Materials required: Collections of gastropod mollusk shells and/or common South Carolinian fishes with alphanumeric labels, dichotomous keys (provided in lab manual)

Directions

1. Working in pairs, you will identify the different species of gastropod mollusks and fishes (if available) common to the South Carolina coast. Each taxon has its own key and specimens, and the specimens are only identified by numbers.

2. Use the dichotomous keys below to assign the scientific names to these species in the table provided on the following pages. For full credit, you must include the genus and species name in the correct format for every organism.

3. Some helpful advice on navigating a dichotomous key:

    a. Before you start working your way through the key, look at the diagram of the organism and familiarize yourself with the labeled parts so you can correctly identify each of the features on the organism. Dichotomous keys are filled with taxonomic and anatomic language which can be difficult to understand if you are not familiar with the terminology.

    b. With each new organism, always start at the beginning of the key. The choices are always sequential; that is, they result from each other in a specific order, and are numbered in this manner. Starting with line number one, read the two character state choices (A and B). Once you have made your choice—either A or B—you will notice that the key gives you options, listed either as a number or as a species. If you get a number, that directs you where to go next in the key. If you get a species, you've got the answer!

GASTROPOD MOLLUSKS

Gastropod means "stomach-foot," meaning that the foot and stomach were located in the same place in this group's common ancestor. Gastropods are characterized by having (at some point in their development) a single shell, as compared to two shells such as in oysters and clams. Figure 5.3 provides a diagram of the general characteristics of a gastropod mollusk.

FIGURE 5.3

General diagram of a gastropod mollusk with major morphologic features labeled for use in identification.

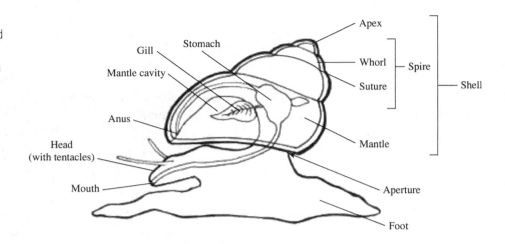

Source: Rachel E. G. Kalisperis

KEY TO SOME COMMON GASTROPOD MOLLUSKS

1   A. Shell not coiled (torted) (i.e., no spire)..2

      B. Shell coiled (torted)...3

2(1) A. Shell shaped like a small low dome .... Cayenne Keyhole Limpet
                                                                           (*Diodora cayensensis*)

      B. Shell capped with a small pocket or shelf underneath................Atlantic Slipper Shell
                                                                           (*Crepidula fornicata*)

3(1) A. Aperture (mouth) rounded or elliptical..4

      B. Aperture (mouth) elongate...6

4(3) A. Shell length and height nearly equal 5

      B. Shell compressed—shell width much greater than height .... Baby's Ear
                                                                      (*Sinum perspectivum*)

5(4) A. "Top" shaped, thick-ridged shell, small < 4 cm.................................Marsh Periwinkle
                                                                      (*Littorina irrorata*)

      B. Globular (ball) shaped shell, smooth surface............................... Atlantic Moon Snail
                                                                     (*Neverita duplicata*)

6(3) A. Exterior shell surface rough...7

      B. Exterior shell surface smooth, with cream background and brown markings,

         aperture not along entire shell .... Banded Tulip (*Fasciolaria hunteria*)

7(6)   A. Pear-shaped, with large, heavy whorls (shell spires) on "shoulder" of shell 8

       B. Spindle-shaped, whorls not shouldered, < 7 cm 9

8(7)   A. Thick-shelled with large knobs on whorl shoulder... Knobbed Whelk
                                                                                                (Busycon carica)

       B. Thin-shelled without large knobs on shoulder, groove around spire.................Channeled Whelk
                                                                                          (Busycotypus canaliculatus)

9(7)   A. Small, black-brown color, no prominent ridges on shell.............................Common Mud Snail
                                                                                              (Ilynassa obsoleta)

       B. Gray color, ridges, aperture pointed...Atlantic Oyster Drill
                                                                                             (*Urosalpinx cincerea*)

ESTUARINE FISHES

Figure 5.4 provides a generalized diagram of the major identifying features of fish that are used in the dichotomous key. Refer to the figure to locate the features as you work through the specimens.

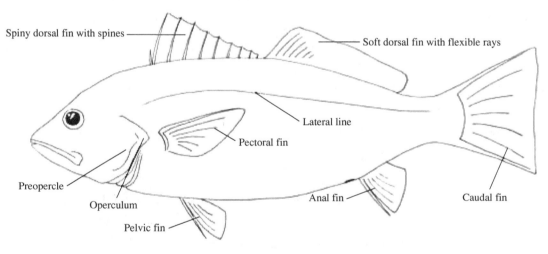

FIGURE 5.4
General diagram of an estuarine fish with major morphologic features labeled for use in identification.

Source: Rachel E. G. Kalisperis

KEY TO SOME COMMON SOUTH CAROLINA ESTUARINE FISHES

1   A. Pelvic fins present, abdominal (i.e., closer to tail) 2

       B. Pelvic fins present, attached to the thorax or throat; under, anterior to or slightly
           behind, base of pectorals.. 4

2(1)   A. Dorsal fin spines and soft fin rays are continuous 3

       B. Dorsal fin spines and rays are separate .... Striped mullet *(Mugil cephalus)*

3(2)   A. Dorsal fin rays number 11 .... Mummichog *(Fundulus heteroclitus)*

       B. Dorsal fin rays number between 13 and 15 .... Striped Killifish *(Fundulus majalis)*

4(1)   A. Dorsal fin spines and rays are continuous...5

       B. Dorsal fin spines and rays are separate 7

5(4)   A. Form symmetrical, eyes and color are not confined to one side 6

       B. Form asymmetrical, eyes and color only on one side, leaving the other side blind and

          colorless..9

6(5)   A. Teeth are small, preopercle smooth-edged ....Pigfish *(Orthopristis chrysoptera)*

       B. Teeth are pronounced, preopercle serrate .... Pinfish *(Lagadon rhomboides)*

7(4)   A. No barbels on lower jaw...8

       B. Short barbels present on lower jaw............................ Atlantic Croaker *(Micropogonias undulatus)*

8(7)   A. Teeth very small .... Spot *(Leiostomus xanthurus)*

       B. Teeth are pronounced (may be difficult to see).........................Silver Perch *(Bairdiella chrysoura)*

9(5)   A. Eyes are large, usually widely separated..10

       B. Body elongate; eyes are small .... Blackcheek Tonguefish *(Symphurus plagiusa)*

10(9) A. Lateral line on ocular side (side with eyes) strongly arched above pectoral fin.............................

         ....Summer Flounder *(Paralichthys dentatus)*

       B. Lateral line on ocular side weakly arched above pectoral fin .... Fringed Flounder

                                                               (Etropus crossotus)

GASTROPOD MOLLUSKS

Number	Common Name	Latin Name
1		
2		
3		
4		
5		
6		
7		
8		
9		
10		

ESTUARINE FISHES

Number	Common Name	Latin Name
1		
2		
3		
4		
5		
6		
7		
8		
9		
10		
11		

EXERCISE 5.2
THE "NUTS AND BOLTS" OF TAXONOMY

Materials required: Bag of various hardware (nails, staples, screws, etc.), blank sheets of paper, tape, colored pencils, rulers, hardware identification sheets

Directions

As renowned taxonomists, your group's goal is to develop a classification scheme and phylogenetic tree of the hardware "organisms" in front of you that meets the established rules of the Linnaean System as well as reflects "evolutionary" trends in your organisms. Follow the guidelines below to help you develop your tree and classification.

PART 1: MAKING A PHYLOGENETIC TREE

1. Your first task is to determine the characters for your organisms. Examine all the hardware and look for similarities and differences among your organisms to determine what features are characters. You should think about what role the form and function of this character has played in the evolution of these animals, and how this character might have evolved. Examples of character traits include "Flat head vs. Rounded head," "Pointed (screw) tip vs. Flat (bolt) tip," and "Steel vs. Iron vs. Galvanized steel" metals.

2. Next, consider which characters or traits are plesiomorphic (ancestral) or apomorphic (derived) and how these traits may have evolved through the lineage—what is the "direction" of character evolution? Which is ancestral, and which is derived? If ancestral, what did it evolve into (if you think it has done so)? If derived, did it derive from another character for a particular function? You should be able to provide a reason for the evolution of each new characteristic that arises in your tree.

3. Now, place your organisms in a logical, coherent phylogenetic tree containing all the pieces of your hardware organisms, based on the evolution of the characters and when the different synapomorphies arose.

4. Draw your phylogenetic tree in the space provided on Page 85.

PART 2:   DEVELOPING A CLASSIFICATION SCHEME

1.   From your phylogenetic tree, you will now develop a classification scheme (the taxonomy) for your organisms (see Table 5.1). You *must* include the following major categories: Kingdom, Phylum, Class, Order, Family, Genus, species.

2.   Decide in your group which of the characters are phylum-level, class-level, order, family, genus, or species level. To determine this, use the following considerations:

   a.   Based on its form or function, and relative to other traits the organisms possess, would a particular character be considered a phylum, class, order, or other level of classification? What should you use to distinguish these different categories? One way to determine this is to think about what classifies all of one lineage (branch) of the hardware organisms tree, is it head shape, threading, metal type? For example, if all your organisms fall into three separate metal types, would this be a class or a phylum?

   b.   Then, what is the next character trait evolution that would allow separation of the hardware organisms within that category you just identified into different subcategories? If you had considered all metal types as separate phyla, then within each phylum, the next category separating your organisms would be class-level.

   c.   You must be able to explain and justify what characters you used to separate each category in your classification. In other words, you should be able to explain the evolutionary or **phylogenetic** relationship among the characters (because this is a phylogenetic tree, not a taxonomic or dichotomous tree).

3.   Provide descriptive names for each of your categories of phylum, class, order, family, genus, and species. Apply names that best describe each object and their hierarchical location in your classification scheme. Be creative when making up the "Latinized" names of your objects! In the Linnaean classification system, typical endings for the taxa are (but not restricted to):

Class or Subclass	-edia, -ia, -ida, -idea, -oda
Order or Suborder	-acea, -aria, -iata, -inea, -oidea
Family or Subfamily	-aceae, -idea, -inae

4.   Place the taxa and species names for your organisms in their correct location on the tree. From this, now write out your Classification Scheme for your organisms, which lists each taxon and its respective subcategories, including species names of your organisms. Do this on a new sheet of paper.

5.   Based on your development of a phylogenetic tree and classification scheme, answer the following questions:

   a.   What are some of the similarities and differences between classifying inanimate objects and living organisms? Provide two examples for each:

   Similarities: _____

   _____

Differences: _____

_____

b. Is it easier or more difficult to classify living or inanimate objects? Why?

_____

_____

_____

c. If you knew nothing about the function of a particular characteristic (the threading on a bolt, the shell of a snail), would that make a difference in the classification scheme developed for that particular group of organisms? Why or why not?

_____

_____

_____

d. On the blank, back side of this page, write a paragraph describing the evolutionary relationships among your hardware organisms, based on your phylogenetic tree and classification scheme. Include in this description the following information:

- What groups (or characters) evolved from whom?
- What natural selection factors (environmental, predation, etc.) might have caused this "speciation"?
- Which shape do you consider to be the most primitive, or plesiomorphic, and why?
- Which do you consider the most advanced, or apomorphic, and why?

Additional Information

If you are interested in further information about the topics in this lab, the following sources will provide you with more descriptions on taxonomy, phylogenetics, and systematics:

▸ Blackwelder, Richard E. 1967. *Taxonomy; a text and reference book*. New York: John Wiley & Sons, Inc. 698 pp.
▸ Wiley, E. O. 1981. *Phylogenetics: The Theory and Practice of Phylogenetic Systematics*. New York: John Wiley & Sons, Inc.

Question 5.d.: in a couple of paragraphs, describe the evolutionary relationships among your hardware organisms, based on your phylogenetic tree and classification scheme.

PHYLOGENETIC TREE

(Draw)

Organism Number	Kingdom	Phylum	Class	Order	Family	Genus	Species

LAB 6

FOOD WEBS AND TROPHIC LEVELS

OBJECTIVES

- ▶ Investigate the interactions between food webs and trophic levels in a typical South Carolina estuary.
- ▶ Apply research tools and data to learn important productivity concepts.
- ▶ Learn the concepts and terminology used in describing ecosystem trophic dynamics.

POINTS TO PONDER

- ▶ Why do the most productive regions have the shortest food chains?
- ▶ Why are fisheries more abundant along coasts?
- ▶ Why does the open ocean have relatively few large organisms?

TERMS TO KNOW

biomass	estuary	nutrient regeneration
carnivore	food chain	particulate organic matter
decomposition	food web	primary consumer
detritivore	herbivore	primary producer
detritus	heterotrophic	secondary consumer
dissolved organic matter	microbial loop	standing stock
ecosystem	net primary production	trophic level

WHAT IS AN ECOSYSTEM?

All organisms are part of a larger dynamic environment, called the ecosystem. An **ecosystem** is the total chemical, physical, geological, and biological components of the environment, from the tiniest algal cell to the sand grains on the seafloor. Marine ecosystems differ widely in size, complexity, and diversity of organisms, but they all share some common characteristics.

Ecosystems are governed by the First Law of Thermodynamics, which states that the total amount of energy in a system is conserved, i.e., matter is neither created nor destroyed. While the *kind* of energy in a given system can change (from solar energy to biochemical energy, to heat energy), the *total* amount cannot. All of Earth's systems, both living and nonliving, transform the sun's energy into some other form. The concept of the **food chain** and its trophic levels is a particularly useful way of tracking the changes of energy as it flows through living systems in an ecosystem. A **trophic level** consists of all organisms that obtain their energy from the same source, no matter their size or classification. The first trophic level in most ecosystems contains the plants (or algae), the **primary producers** (Figure 6.1). As you learned in the Primary Production lab, organisms at this level perform photosynthesis, absorbing solar radiation to drive chemical reactions that make plant tissues and other complex molecules. These molecules, such as glucose and other sugars, are subsequently used as energy sources by organisms in higher trophic levels. The second trophic level includes the **herbivores**, animals that get their energy by feeding directly on the primary producers. The herbivores are therefore called the **primary consumers**. Organisms at this trophic level are **heterotrophic**, meaning that they do not make their own energy directly from the sun but consume it from others. The next, or third, trophic level consists of **carnivores**, those animals that get their energy by eating other heterotrophic organisms, the herbivores (or other carnivores). These organisms, while also heterotrophic, are called the **secondary consumers**. At the end of the food chain are the top predators, such as killer whales.

The number of trophic levels in an ecosystem primarily depends on location and the size of the individual primary producers. The open ocean ecosystem can contain as many as six trophic levels, where tiny photosynthetic primary producers, the nano-

FIGURE 6.1

Schematic diagram of a simple food chain in the open ocean illustrating four trophic levels.

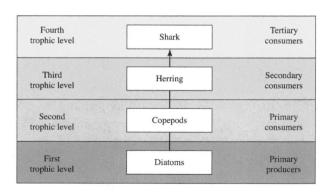

Source: Michelle L. Hardee

plankton (10^{-9} m), are fed upon by successively larger trophic levels of zooplankton and fish. The shortest trophic levels are in upwelling and polar regions, where large, chain-forming diatoms (single-celled algae primary producers) are fed upon directly by fish or krill (a type of crustacean), which are in turn fed upon by baleen whales. Most ecosystems are much more complex and have several different primary producers, with animals consuming more than just one kind of food. Many animals will shift the type of food (and trophic level) they consume as they age. The size of the individual organisms generally increases within each succeeding trophic level, but the length of the life cycle also becomes longer. Thus, the simplistic straight-line food chain shown in Figure 6.1 can be more accurately referred to as a complex **food web** (Figure 6.2).

When discussing ecosystems and food webs, scientists prefer to use terms that refer to the amount of material or energy present, not necessarily the organisms themselves. The purpose of studying these trophic dynamics is to quantify the amount of *energy* being transferred between different trophic levels, and the amount being lost as heat, animal tissue, decomposing material, and so forth. The majority of primary production in the oceans is accomplished by small, open ocean phytoplankton rather than marine plants such as seaweed or salt marsh grasses (plants only account for 5–10% of total marine productivity). However, you should recall that the primary producer uses a portion of this organic material produced by photosynthesis for metabolic functions, e.g. respiration. The excess, or **net primary production,** is the amount of organic material *available to support higher trophic levels*: primary and secondary consumers and decomposers. Net primary production is what is available to support the rest of the ecosystem.

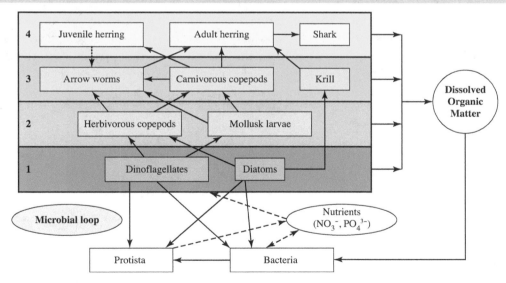

FIGURE 6.2

Schematic diagram of a food web, including the microbial loop. Feeding relationships exhibited by organisms in an ecosystem are understood by examining the network of interconnected and interdependent food chains. Numbers represent trophic "levels" as identified in Figure 6.1. Arrows point towards the direction of energy transfer, i.e., from a food source *towards* its consumer. The dotted arrow represents predation of a lower trophic level on a higher level. The dashed arrows represent excretion of nutrients which are then taken up by all phytoplankton. See text for discussion of the microbial loop.

Source: Michelle L. Hardee

Surprisingly, much of the energy contained in a particular trophic level is consumed almost entirely by the activities of the organisms in *that level*. Energy and organic matter are also lost as waste materials. Only about 5–20% of the energy is actually passed from one level to the next, depending on the ecosystem. The average amount of energy transfer is approximately 10%. A large portion of the material is lost from the trophic level through sloppy feeding, cell leakage, waste products, and dead organisms, collectively called **detritus**. This material is known as either **dissolved organic matter** (DOM) or **particulate organic matter** (POM), depending on the size of the particle. The energy contained in this material is not lost to the ecosystem, however. Bacteria, fungi, and other organisms that perform **decomposition** break down this non-living organic matter into its original components of CO_2, water, and nutrients. These organisms are **detritivores**, consuming detrital material, and thus serving a vital function in the ecosystem: to recycle or release nutrients that were incorporated into the organic matter during primary production, making those nutrients again available to photosynthetic organisms. This process of **nutrient regeneration** is a fundamental component of all ecosystems. It is part of the **microbial loop**, the portion of the food web where nutrients are regenerated and returned to the system by microorganisms such as protozoans and bacteria. The single-celled protozoans are an important consumer of bacteria and play a large role in the recycling of nutrients within the food web.

Other terms are used to refer to the amount of material represented by the organisms. The **standing stock** refers to the number of organisms per unit area or per unit volume of water at the moment of sampling. While we can count numbers of phytoplankton from filtered seawater samples, this method is time-consuming. Instead, we use chlorophyll *a* concentrations, the photosynthetic pigment used by all primary producers. This method is more accurate because the ability of chlorophyll *a* to fluoresce when excited by an appropriate wavelength of light can be easily measured. The **biomass** is defined as the total weight (*total numbers × average weight*) of all organisms in a given area or volume. This measure is used in preference to standing stock because phytoplankton vary in size, and the total numbers of all the organisms is not as ecologically meaningful as are estimates of their biomass.

BACKGROUND ON THE EXERCISES

North Inlet, an estuary located north of Georgetown, S.C., is a National Estuarine Research Reserve (NERR). **Estuaries** are regions where saltwater and freshwater meet, typically at the mouth of a river or an inlet or bay. North Inlet is home to research focusing on many aspects of the estuary, including its resident population of bottlenose dolphins, *Tursiops truncatus*. Marine mammals are believed to be very important to the structure and function of marine communities. As top predators, they take a significant amount of prey from any given system, and in turn, they are affected by changes in the system. These dolphins can thus serve as an indicator species for seasonal and long-term changes in the estuary. The dolphins in North Inlet spend most of their time within the tidal salt marsh creeks of this system, an area 32 km², though they also use the adjacent waters of Winyah Bay and the Atlantic Ocean (Figure 6.3 on following page). Studies such as this provide data for future management decisions concerning marine mammals, as well as the fisheries represented by their prey.

Much of the research depends upon the ability to identify and track the dolphins in North Inlet. To do this, "focal follow surveys" are conducted, in which a specific dolphin or group of dolphins are followed continuously as they go about their daily lives in the estuary. Dolphins are photographed during each study day and are then identified by the unique shape of their dorsal fins. Trying to match pictures of fins with a "catalog" of previously identified fins is challenging, and requires researchers to develop an eye for detail. Photo-identification techniques have become standard practice for cetacean biologists. The natural marking on the dorsal fins, including notches, nicks, and scars, are used for identification.

Building on previous ecological research by many other scientists, Young and Phillips (2002) developed an ecological model to estimate the impact of the resident dolphin population on the North Inlet food web. This activity examines this model in detail and consists of multiple exercises. The first requires you to practice the challenge of dolphin identification using actual photos of North Inlet dolphins. The subsequent exercises investigate the dolphins' role in the food web by step-wise calculation of the annual primary production required to support them.

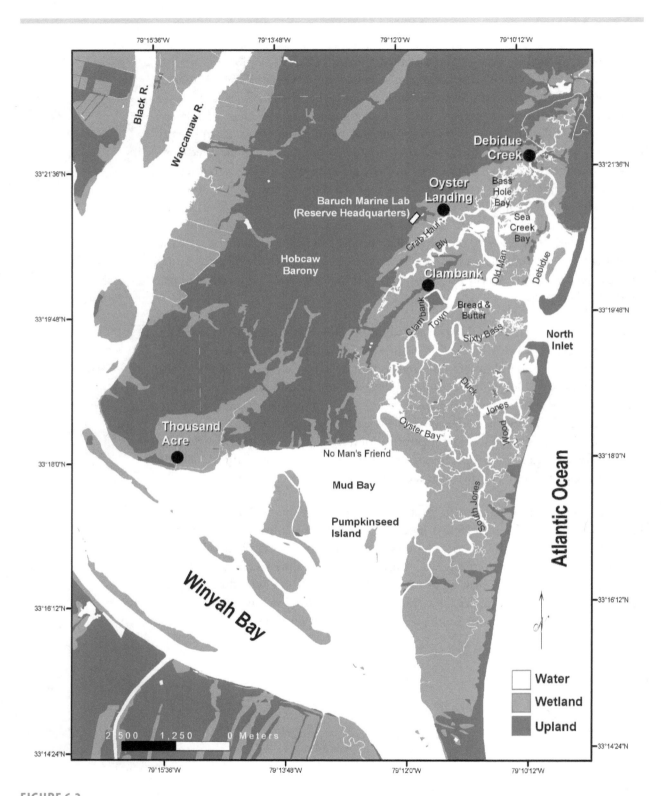

FIGURE 6.3

Location of North Inlet National Estuarine Research Reserve and Winyah Bay. Map courtesy North Inlet-Winyah Bay National Estuarine Research Reserve (NI-WB NERR), Baruch Marine Field Lab.

EXERCISE 6.1
DOLPHIN PHOTO-IDENTIFICATION

Materials needed: Catalogs of named and unlabeled photos of North Inlet dolphin population dorsal fins.

OBJECTIVE

Your goal is to correctly identify individual dolphins in North Inlet and Winyah Bay based on the distinctive characteristics and natural markings of their dorsal fins, as well as identifying some basic patterns of association and movements, using photo-identification. This is a proven and applied research tool used by scientists to track the movements and associations of dolphins in different ecosystems.

PART 1: PHOTO-IDENTIFICATION OF INDIVIDUAL DOLPHINS

1. Your instructor will provide you with a catalog of 10 fin photographs, each representing a different individual dolphin in North Inlet. The catalog is followed by 25 *unlabeled* fins. It is your job to match each unlabeled fin with one of the known fins in the catalog.

2. Some considerations:
    a. To identify individuals, researchers pay close attention to the shape and placement of notches along the perimeter of the fin, as well as the overall shape and curvature of the fin. Fin notches can be caused by sharks, boat strikes, rope or line entanglements, other dolphins, etc.

    b. Surface scratches can last for months and are a useful identification feature for the short term. Most of these temporary marks are tooth rakes from other dolphins.

    c. Mother and calf pairs tend to always be seen together.

    d. The color of the fins is not a factor in identification, as color can vary with camera type, clouds, etc.

    e. In the data table on the following page, an "event" is an occurrence on a particular day when a dolphin or group of dolphins was seen. If dolphins were seen on the same day *and* the same event, they were part of the same group on that day and therefore are considered to associate together.

3. As you identify each unlabeled fin, place the names of the dolphins in the chart below. You will find that some fins are quite obvious, while others are more difficult. Photo-identification is a very useful tool, but it takes some practice, so don't get discouraged if you can't immediately identify individual dolphins.

4. When you are done, answer the questions following the table.

TABLE 6.1

Photo-identification of individual dolphins in North Inlet.

Picture	Name	Date of Picture	Event Number
1		2/16/2002	1
2		6/9/1999	1
3		3/14/2002	2
4		3/14/2002	2
5		5/31/2002	4
6		5/31/2002	3
7		5/21/1999	2
8		5/9/2002	3
9		5/22/1998	1
10		5/29/2002	1
11		5/31/2002	4
12		5/17/2002	1
13		5/9/2002	2
14		6/5/2002	2
15		3/26/2002	4
16		3/26/2002	3
17		3/26/2002	2
18		5/31/2002	5
19		5/31/2002	5
20		5/31/2002	3
21		5/17/2002	1
22		5/23/2002	1
23		5/23/2002	1
24		6/4/1998	1
25		4/13/1999	1

PART 2: PHOTO-IDENTIFICATION ANALYSIS

1. According to the data in the table on the previous page, which dolphins associate together (i.e., are in the same group)?

   _____

   _____

2. What mother and calf pairs are present (almost always seen together)?

   _____

3. Are there any dolphins that do not seem to interact with others (never seen together, even if photographed on the same day)?

   _____

   _____

4. Killer whales are typically found in consistent groups of multiple family members. Does your data indicate that bottlenose dolphins have a similar social structure?

   _____

   Explain. _____

   _____

From this exercise, you should have determined some basic patterns of association between dolphins. The same dolphins are found in North Inlet over the course of months and even years. In fact, North Inlet supports a resident group of dolphins that spend most of their time within the region throughout the year. Though it is beyond the scope of this activity, photo-ID and focal follow studies have identified daily and seasonal movement patterns within the creek system and have helped to determine the average number of dolphins that utilize the area. These details are important in order to understand the ecological role of dolphins, the subject of the next exercise.

EXERCISE 6.2
THE ECOLOGICAL ROLE OF BOTTLENOSE DOLPHINS, *TURSIOPS TRUNCATUS*, IN AN ESTUARINE FOOD WEB

Materials needed: Calculator, pencils

OBJECTIVE

In this exercise, you will determine the ecological role and food web interactions affecting the bottlenose dolphin population in the South Carolina North Inlet NERR salt marsh food web ecosystem.

PART 1: CONSTRUCTING A FOOD WEB OF THE NORTH INLET ECOSYSTEM

Based on studies of stranded animals and a study of North Inlet fishes, the table below lists likely candidates for dolphin food sources:

TABLE 6.2

Food sources for *Tursiops truncatus* in the North Inlet food web.

Common Name	Scientific Name	What it eats	Trophic Level
Red drum	*Sciaenops ocellatus*	Crustaceans & fish	4
Speckled trout	*Cynoscion nebulosus*	Fish	4
Weakfish	*Cynoscion regalis*	Crustaceans & fish	4
Striped mullet	*Mugil cephalus*	Detritivores (plant material, bacteria, & invertebrates)	2
Flounders	*Paralichthys* spp.	Crustaceans & fish	4
Silver perch	*Bairdiella chrysoura*	Crustaceans, polychaetes, & nematodes	3
Spot	*Leiostomus xanthurus*	Benthic invertebrates	3
Croaker	*Micropogonias undulatus*	Crustaceans, mollusks, fish, & polychaetes	3
Pinfish	*Lagodon rhomboides*	Benthic invertebrates	3
Black drum	*Pogonias cromis*	Mollusks, arthropods, annelids, & fish	3
Ladyfish	*Elops saurus*	Crustaceans & fish	4
Toadfish	*Opsanus tau*	Crustaceans & fish	4
Eels	Various species	Crustaceans & fish	4
Shrimp	Various species	Detritivores	2

1. What does a simplified version of the North Inlet food web look like? Using the trophic level information in the table above and what you now know about trophic food webs and estuaries, sketch a simplified North Inlet food web below. Start with the first trophic level, and end with dolphins as the top predator. You may assume that the bulk of their diet (~95%) is fish.

PART 2: ESTIMATING A REPRESENTATIVE TROPHIC LEVEL FOR THE BOTTLENOSE DOLPHINS

1. The next step consists of estimating the average trophic level represented by the dolphins in the North Inlet system. Look at your food web and the trophic level information presented for the prey. Your task is to come up with one number (it can be a decimal number) to represent the trophic level of bottlenose dolphins in North Inlet, based on the trophic levels they feed upon. Discuss in your group the different contributions of prey to the dolphin trophic levels (how much of each prey item do the dolphins eat?), and how to determine such a number.

2. What is your estimate for the trophic level of dolphins in North Inlet? _____ How did you come up with this number? Show your calculations below:

3. This seems like a round-about way to figure out what a dolphin eats. How is diet determined in studies of other animals?

_____

_____

_____

4. Why is this more difficult with dolphins?

_____

_____

_____

EXERCISE 6.3
CALCULATING THE PERCENTAGE OF PRIMARY PRODUCTION REQUIRED TO SUPPORT *T. TRUNCATUS* IN A NORTH INLET FOOD WEB

OBJECTIVE

In this exercise, you will determine both how much prey and associated primary production are necessary to support the small dolphin population in the North Inlet system. The calculations aren't hard, but they require you to concentrate, think, and work together.

PART 1: HOW MUCH PREY IS AVAILABLE TO THE DOLPHIN POPULATION?

In Part 1 of this exercise, you will figure out how much prey is available to the dolphin population in North Inlet. If you know the total amount of primary production in North Inlet, and you know what proportion of energy is transferred from one trophic level to the next, you can calculate how much prey should be available at a given trophic level. However, there are several things to first consider. Primary production in the North Inlet system may be broken down into four categories based on the groups that produce it: phytoplankton, benthic microalgae, *Spartina* species, and macroalgae. The first two (phytoplankton & benthic microalgae) are grazed on *directly*, thus providing a more efficient path for energy to proceed up the food chain. The second two (*Spartina* species and macroalgae) are not fed upon directly. Instead, they "leak" organic carbon as dissolved or particulate organic matter (DOM and POM), or they die and are broken down by bacteria, and small plankton graze upon the bacteria. This is the **detrital** or **microbial loop,** and must be factored into any production calculations. For our model, we have added *1.5* trophic levels to these paths to represent the extra steps. The estimated annual primary production (based on available data) in North Inlet is shown in Table 6.2.

TABLE 6.2
Estimated annual primary production in North Inlet. P = Annual production at that particular trophic level, expressed as minimum, maximum, and percentage estimates.

	Producer Category	P (metric tons C / yr) *		P (%)
		Minimum Estimate	Maximum Estimate	
Direct Sources	Phytoplankton	1,370	2,770	15–26
	Benthic Microalgae	3,423	3,423	22–38
Indirect Sources	*Spartina* spp.	2,730	7,900	30–59
	Macroalgae	1,537	1,537	10–17
	Total	9,060	15,630	100

* Note that 1 metric ton = 1×10^6 grams.

Once you calculate the annual primary production, you will then calculate the available prey (essentially the amount of carbon being consumed) using the following formula:

$$C = P \times E^{(TL-1)}$$

where:

P = the annual primary production (from Table 6.2)

E = the transfer efficiency between trophic levels, expressed as a decimal proportion (not a percent). The transfer efficiency E is defined as:

$$E = \frac{P_t}{P_{(t-1)}} \cong 0.1$$

where:

P_t is the annual production at trophic level t

P_{t-1} is the annual production in the preceding trophic level $t-1$

Traditionally, ecological models have assumed an energy transfer rate of 10% or 0.1 for trophic transfer rates and that rate is what will be used in this lab.

TL = the average prey trophic level

C = the resulting annual production of prey (food available for the whole year)

For your calculations, you will use *both* the minimum and maximum estimates for primary production in Table 6.2 above to calculate a minimum and maximum estimate for C. It is common to calculate a range of estimates in ecological models.

1. Look at the trophic level you calculated for the dolphins in Exercise 6.2. What is the average trophic level of their prey (TL)?

    _____

    _____

2. Based on the definition of E above, which trophic levels could you use to find E and why?

    _____

    _____

3. Why are most determinations of E approximately 10%, not higher? _____

    _____

4. Calculate the maximum and minimum estimate for *C*, the resulting annual production of prey. In your calculations, you will need to treat the direct and indirect sources of primary production separately.

- For the **direct** sources—phytoplankton and benthic microalgae—assume the prey are supported at the trophic level given in your answer to Question 1 above. Calculate both a minimum and maximum estimate of *C* for each source and sum the two sources for your total primary production minimum and maximum from direct sources.

- For **indirect** sources—*Spartina* spp. and macroalgae—you must first *add 1.5 trophic* links to your prey *TL* calculations to account for the microbial loop. Calculate both a minimum and maximum *C* for each source, and sum for your total min/max from indirect sources.

- Combined, your results from both direct and indirect sources should add up to a minimum and maximum total annual production of prey from all sources.

Show your calculations here (remember to include units!):

Direct Sources   **Indirect Sources**

The range (minimum and maximum) of prey production values for the dolphins from all sources:

Minimum estimate: _____

Maximum estimate: _____

PART 2: ESTIMATING THE TOTAL ANNUAL PREY CONSUMPTION
OF NORTH INLET'S BOTTLENOSE DOLPHIN POPULATION

This next step requires you to estimate the total annual prey consumption of North Inlet's bottlenose dolphin population. Consider the following information (you will use *all* of it in the following calculations):

1.	A single dolphin eats an average of 6.4 kg of food per day (for simplicity, you may assume it is all fish).
2.	The energy available from fish can be estimated using 1,440 kcal/kg fish.
3.	Caloric value can be converted to grams carbon using 10 kcal/g C.
4.	The average number of dolphins observed in North Inlet per day is 8, but they only spend, on average, about 75% of their time within the North Inlet system.
5.	1 metric ton of Carbon = 10^6 grams of Carbon

1. How much does a *single dolphin* eat per day, expressed as grams of Carbon? Show your calculations here:

2. How much does the North Inlet dolphin population eat per year, again expressed in g of C? Show your calculations here:

PART 3: WHAT PERCENTAGE OF PRIMARY PRODUCTION IS REQUIRED TO SUPPORT THE NORTH INLET DOLPHIN POPULATION?

In Part 1, you calculated the annual available prey based on primary production in North Inlet. In Part 2, you calculated the annual consumption of the dolphin population in North Inlet. Now, let's conclude our investigation of the role of *Tursiops truncatus* in the North Inlet food web by determining the percentage of primary production required to support these dolphins.

1. What *percentage* of the total annual primary production available in North Inlet is required to support the dolphin population? Show your calculations here: (You should have a maximum and minimum estimate.)

2. What percentage is required to support just one dolphin? _____ Show your calculations here:

DISCUSSION

1. Would you consider the amount of primary production required to support a single dolphin in North Inlet to be "a lot"? Defend your answer (are you considering it in terms of %, 'C', or what?).

_____

_____

2. How many other species are also supported by the same primary production? (Look at the prey table.)

_____

3. Do you think it takes a comparable amount of energy to support *you*? Why or why not? (Hint: think about where your food comes from, i.e., what does it take to produce, harvest, package, and get the food to your table?)

_____

_____

_____

LAB 7

SURVEY OF MARINE ORGANISMS
PART I

▶ Learn the identifying characteristics of the major marine phyla of invertebrate organisms and be able to recognize these groups and some of their subgroups.

▶ Be able to identify general evolutionary trends and how these apply to the marine environment.

POINTS TO PONDER

▶ Why is a sponge considered an animal?
▶ What is a coelom and why is it important in the evolution of more advanced organisms?
▶ What is the purpose for having a segmented body?
▶ Why are there so many different phyla of worms in the Animal Kingdom?

TERMS TO KNOW

Acanthocephala	choanocyte	Ctenophora	hermaphroditic
acoelomate	Cnidaria	dorsoventral	incurrent pore
alternation of	cnidocyte	ectoderm	Kinorhyncha
generations	coelom	endoderm	leuconoid
anterior	complete digestive	eucoelomate	medusa
asconoid	tract	Gastrotricha	mesenchyme
bilateral symmetry	complexity	gastrovascular	mesoderm
caeca	corona	cavity	mesohyl
cephalization	ctene	gonochoristic	monoecious

nematocyst	peritoneum	pseudocoelomate	spongocoel
Nematoda	Platyhelminthes	radial symmetry	syconoid
Nemertea	polyp	rhynchocoel	true circulatory
osculum	Porifera	Rotifera	system
parthenogenesis	posterior	scolex	
pentaradial	proboscis	segmentation	
symmetry	proglottid	spicules	

INTRODUCTION

In the taxonomy lab, you learned how scientists classify organisms, and the logic behind the categorization of organisms. In this lab you will be surveying the major taxa of marine organisms, in order to gain fundamental knowledge of the important phyla in the marine realm. Many students, when asked to name a marine animal, automatically answer "dolphin", "shark", or "fish". As you will discover, very little of the diversity in the marine realm actually lies in such groups. Of the ~32 animal phyla recognized by most taxonomists, all but one consists *entirely* of invertebrate animals, and the remaining one (Phylum Chordata) is partially invertebrates as well!

As you go through this lab, keep in mind that this is only a *survey*. In actuality, there are a vast number of marine organism groups. Some universities have entire courses devoted to the study of individual taxa such as crustaceans and corals. There are too many organisms for us to cover all the aspects of these groups with any reasonable depth here. However, many terms and concepts introduced in these labs will be used by you in every other marine biology course you take.

VARIATIONS ON MANY THEMES

As you study the different animal phyla, you will begin to see patterns, or *themes*, emerge in body plans, shapes, features, and functions. These themes help to illustrate important characteristics underlying the evolutionary relationships between organisms. Such patterns and trends will change as species adapt to their environment, with certain features often disappearing in the more specialized groups. One obvious trend is toward increasing complexity, with organisms that have cells organized into tissues, tissues into organs, and organs into organ systems. Sponges, for example, represent an early stage in the evolution of complexity, while cnidarians (jellyfish, corals, and their relatives) are a slightly later stage. However, such a comparison does not mean that jellyfish evolved from sponges, nor does it mean that jellyfish are *more highly evolved than* sponges! It only means that these organisms' body plans represent the complexity of an ancestral organism that once existed, and these groups have taken different evolutionary paths based on their adaptation to their environment (Figure 7.I.1).

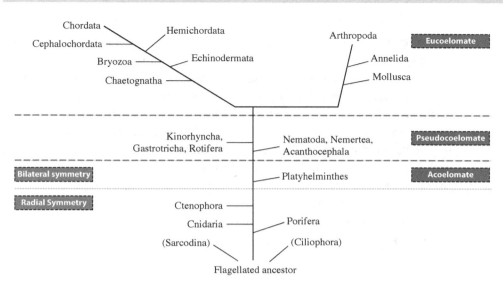

FIGURE 7.I.1

Phylogenetic tree of the major animal phyla. Dashed lines separate phyla by symmetry and types of body cavity. Names in parentheses are not animal phyla but belong to the Kingdom Protista; they are provided as examples of organisms believed to represent possible direct evolutionary lineages from the flagellated ancestor. See text for definition of terms.

Source: Michelle L. Hardee

One of the most important concepts you should learn from this survey of marine organisms is how an organism's body plan has adapted to fit its environment. This concept, which evolutionary scientists refer to by the phrase "form follows function", can be used to understand the adaptations and characteristics seen in organisms. This phrase refers to the result of natural selection, in which an adapted form (characteristic, feature, body shape, etc.) of an organism evolves as a gradual adjustment to constraints imparted by its environment. Such adaptive relationships between an animal's form and its function can be traced through the animal groups as increasingly complex advancements, or trends. As you learn the animal phyla, you should try to identify the six major trends described in the following section.

1. Complexity

As mentioned above, **complexity** refers to the level of organization of an animal's body. A primary characteristic defining animal complexity is the presence or absence of true tissues, which are aggregations of similar cells organized to perform a specific function. Animals such as sponges display a simple level of organization, with only a few types of cells and no true tissues. As organism groups increase in complexity, multiple tissue layers are found, which aggregate into organ systems in more complex animals. Animals such as earthworms display this higher level of organization with true tissues and complete organ systems.

2. Symmetry

This concept refers to the arrangement of parts of an organism into a particular pattern. **Radial symmetry** is similar to a bicycle wheel – identical parts are repetitively arranged in a circle around a central axis. **Pentaradial symmetry** is a special type of symmetry, in which five identical parts are arrayed about a central axis, along five rays of symmetry. It is not true radial symmetry but is derived from bilateral symmetry (see below). Coral polyps exhibit radial symmetry, while most species of sea stars show pentaradial symmetry. More complex organisms have developed **bilateral symmetry,** where the body can be divided into right and left halves that are mirror images of each other. In many cases, this symmetry may be modified, but it is always present in some stage of the organism's life cycle. The development of bilateral symmetry is important because it permits the establishment of **anterior** (front) and **posterior** (back) ends of the body, and cephalization.

3. Cephalization

The concentration of the nervous tissue and sense organs in the anterior end of the body is called **cephalization**. It is thought to have evolved with bilateral symmetry as an adaptation allowing animals to move more safely and efficiently. A bilaterally symmetric animal generally explores the environment by leading with its head, which contains sensory organs used to detect danger, food, etc.

4. Complete digestive tract

Many of the less complex organism groups, like the jellyfish, have no digestive system (an organ system). Instead, they only have a **gastrovascular cavity** with a single opening that functions in both digestion and circulation. Organisms that have both a mouth and an anus have a **complete digestive tract**, such as nematode worms.

5. Development of a body cavity

The development of a fluid-filled cavity in between the body tissue layers, the **coelom**, is a key characteristic for the development and evolution of more complex organisms. The coelom functions as a support structure that provides separation and cushioning of the internal organs from the muscles. This body cavity can arise *only* in organisms that have developed three true tissue layers, the **ectoderm, mesoderm,** and **endoderm**. Among the organisms with three tissue layers, there are three levels of classification (see Figure 7.I.2):

A. **acoelomate**—lacking a true coelom, the only cavity being that of the digestive tract; the mesoderm is essentially a solid mass of tissue between the gut

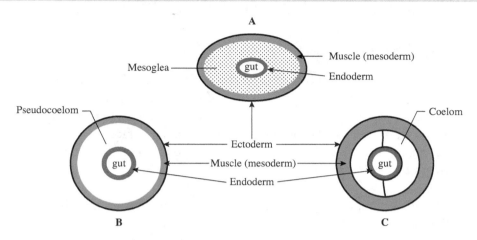

FIGURE 7.1.2

Three types of body cavity conditions found in organisms with three tissue layers. A) Acoelomate. In the acoelomate form the ectoderm is the think black outside line, mesoderm is the solid green line, and endoderm is the solid blue line. B) Pseudocoelomate. In the pseudocoelomate body cavity, the outer solid green line represents mesoderm, and inner solid blue line represents endoderm. C) Eucoelomate. In the eucoelomate form, the coelom arises between two layers of mesoderm (represented by the heavy black lines surrounding the coelomic space), with the solid blue endoderm surrounding the gut. The coelom is shown as a partitioned body cavity in the eucoelomate body plan. .

Source: Michelle L. Hardee

(endoderm) and body wall (ectoderm). Includes the phyla: Platyhelminthes, Gastrotricha.

B. **pseudocoelomate**—having a "false" body cavity, in which the coelom arises as a space between the gut (endoderm) and the muscle (mesoderm). Includes the phyla: Rotifera, Kinorhyncha, Nematoda, Acanthocephala.

C. **eucoelomate**—having a "true" body cavity, in which the coelom arises within the mesoderm itself, and is completely enclosed by a thin lining of mesodermal cells called the **peritoneum**. Includes the phyla: Annelida, Mollusca, Arthropoda, Echinodermata, Chordata.

Within each of these basic body types, animal phyla have evolved a multitude of variations on these themes. Each higher level of complexity has allowed for new variability and adaptation to occur, and enabled organism groups to modify the basic body plans throughout their own particular direction of evolution.

6. Segmentation

The condition in which an organism's body has been divided into multiple separate units that function individually is called **segmentation**. The function of segmentation is to localize muscles for body contraction (such as with earthworms) and for the specialization of repeated parts (such as for reproductive purposes). True segmentation involves the repetition of all organ systems, not just one system.

SURVEY OF THE INVERTEBRATE ANIMAL PHYLA, PART I

Sponges (Phylum Porifera)

Phylum **Porifera** includes the most primitive of multicellular animals, the sponges. Neither true tissues nor organs are found in these animals, and the few cell types present show considerable independence. All sponges are asymmetrical (showing no symmetry); many actually conform to the substrate (surface) to which they are attached. Sponges are also sessile, remaining attached to the substrate and not moving through the environment.

Sponges feed by pumping water through their bodies and filtering out tiny organisms and other particles. Water enters through small **incurrent pores,** and after passing through canals to the central cavity of the sponge (the **spongocoel**), it exits through a large opening called the **osculum** (Figure 7.I.3). The pores, canals, and spongocoel are lined with **choanocytes**, unique flagellated cells with collars. These cells have a flagellum that extends into the opening, moving water through the sponge (Figure 7.I.4). As water passes by, food particles are caught by the flagellum and ingested by the cell. Between the outer and inner cell layers there is a gelatinous layer called the **mesohyl.** This layer contains amoeboid cells (amoebocytes) that secrete **spicules,** skeletal structures that provide support for the sponge body. Sponges are separated into classes according to the type of spicules formed. The spicules are usually composed of either calcium carbonate ($CaCO_3$) or silica (SiO_2). Commercial sponges have a fibrous skeleton composed of a collagen-like substance called spongin.

The simplest sponges are vase-shaped, with an **asconoid** body type (Figure 7.I.5). The body is a thin-walled sac perforated by hundreds of microscopic incurrent pores.

FIGURE 7.I.3

General structures of a sponge body plan. The amoebocytes are found within the mesohyl, and the choanocytes line the incurrent pores and the spongoecoel.

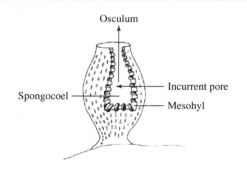

Source: Rachel E. G. Kalisperis

FIGURE 7.I.4

Close-up illustration of choanocytes. The flagellum is surrounded by the collar and extends from the choanocyte into the environment.

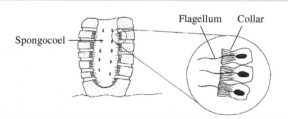

Source: Rachel E. G. Kalisperis

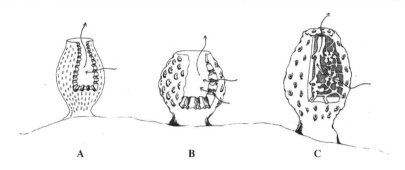

FIGURE 7.1.5

Three body types of sponges: (A) Asconoid, the simplest body type. (B) Syconoid – note the increase in body wall folding in the illustration. (C) Leuconoid, which exhibits complex folding of the body wall and a more irregular shape. The pathway of water in all three types is the same: through the pores and canals, into the spongocoel, and out the osculum.

Source: Rachel E. G. Kalisperis

The **syconoid** body plan is characterized by an increase in the internal surface area for food uptake by the creation of small chambers within the body wall. The greatest amount of surface area occurs in the **leuconoid** sponges. These are the largest and most complex sponges, due to the intricate system of canals and chambers lined by choanocytes. Sponges reproduce asexually by budding and sexually by eggs and sperm. Eggs and sperm are released into the spongocoel at different times. Such sperm release is dramatic, often referred to as "smoking sponges". The egg is fertilized in place and develops into an active larva that escapes from the sponge, eventually settling to become a new organism.

Sponge Taxonomy:

PHYLUM PORIFERA

CLASS CALCAREA: calcareous sponges. Spicules composed of $CaCO_3$. [*Leucosolenia, Scypha, Grantia*]

CLASS HEXACTINELLIDA: glass sponges. Spicules six-rayed, composed of SiO_2. [*Euplectella, Hexactinella*]

CLASS DEMOSPONGIAE: Spicules composed of silica or spongin, commercial sponges. [*Spongilla, Cliona, Microciona*]

Cnidarians (Phylum Cnidaria)

Most of the animals in Phylum **Cnidaria** are marine organisms. They include such forms as the sea anemone, corals, jellyfish, and sea fans. Several important features characterize this phylum. In general, cnidarians are relatively simple animals possessing a basic two-layered body, with a gastrovascular cavity used in the digestion of food and a single opening for both ingestion and excretion. Two body forms characterize cnidarians: the **polyp** form, which is tubular in shape, and the umbrella-shaped

FIGURE 7.I.6

Idealized life cycle of the Cnidaria. Cnidarian classes undergo different pathways depending on which body form is dominant. The medusa and polyp are shown in cross-section to illustrate the 2-layered body plan.

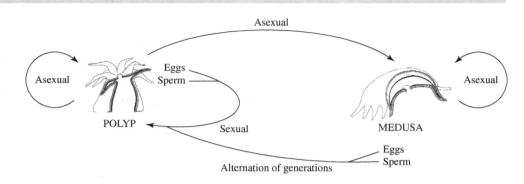

Source: Rachel E. G. Kalisperis

medusa (Figure 7.I.6). The polyp stage is sessile and has the mouth of the gastrovascular cavity directed upward. The medusa stage is free-swimming. Tentacles surround the mouth in both forms. In the Hydrozoan class, there is an **alternation of generations** between the polyp and medusa stages of the life cycle, and the medusae bud off of a specialized polyp. In the other classes, one stage is absent while the other is dominant. The body shape of most Cnidarians exhibits radial symmetry, though this can be modified in many ways.

A unique characteristic of cnidarians is the possession of specialized stinging cells called **cnidocytes** (Figure 7.I.7). These cells contain stinging structures, the **nematocysts**, which are responsible for the painful stings of some jellyfish and corals. When the trigger receives a stimulus (either physical or chemical), the cnidocyte fires its

FIGURE 7.I.7

Process of a nematocyst discharging from a cnidocyte on a hydrozoan tentacle. Note coiling of nematocyst thread within the cnidocyte cell. Illustration: Campbell, A.; Reece, Jane B., *Biology*, 6th Edition, © 2002, p 649. Reprinted by permission of Pearson Education, Inc., Upper Saddle River, NJ.

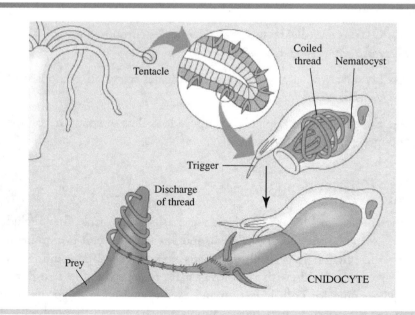

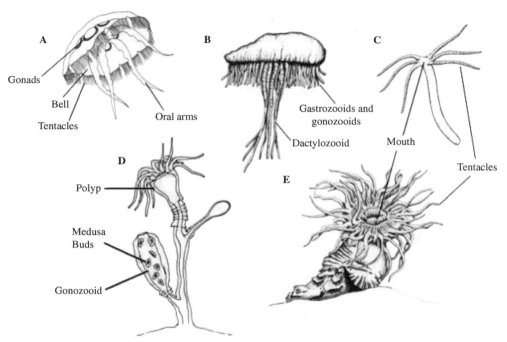

Source: Rachel E. G. Kalisperis

FIGURE 7.1.8

Representatives of some Cnidarian classes. (A) *Aurelia*, the moon jelly, is a scyphozoan. (B) The man-o'-war, *Physalia*, is a hydrozoan. Gastrozooids are modified feeding polyps, dactylozooids are nonfeeding fishing tentacles, and the gonozooids produce reproductive medusae (which are never released). The float is a separate polyp as well. (C) *Hydra*, the common hydroid. (D) *Obelia* is a hydrozoan in which the adult displays both a polyp and a medusa form. (E) The familiar sea anemone belongs to Class Anthozoa. Illustrations are not to scale.

nematocyst, which may either stick to the prey, entangle the object, or deliver a toxin, depending on the type of nematocyst.

Cnidarians are classified according to the primary body form present (Figure 7.1.8). Hydrozoans have both polyps and medusae present at some point in their life cycle. Scyphozoans (true jellyfish) only have medusae, while Anthozoans (corals) have only the polyp present.

Hydrozoans are common members of shallow tropical environments such as coral reefs and sea grass beds, growing in clusters of branched, feathery colonies. A well-recognized but often incorrectly classified hydrozoan is the man-o'-war, *Physalia physalis*. *Physalia* is actually a colony containing many different polyps, all with separate functions (reproduction, prey capture, flotation), but living as a single, colonial organism. Scyphozoans are the true jellyfish, such as the cannonball jelly (*Stomolophus*) or moon jelly (*Aurelia*) that typically wash up along the southeastern U.S. coastline. Anthozoans include the hard and soft corals of coral reefs, as well as the sea fans and sea whips that are also commonly found on beaches. These are colonial animals, with numerous polyps secreting a hard carbonate or keratinized "skeleton" for support. Another member of the anthozoans is the sea anemone, a solitary polyp found on jetties at the beach and often seen in saltwater aquaria. When disturbed, sea anemones will shoot out their filamentous guts as a defense against predators. *Metridium* is a sea anemone typically used for demonstration of the anthozoan body type (Figure 7.1.9).

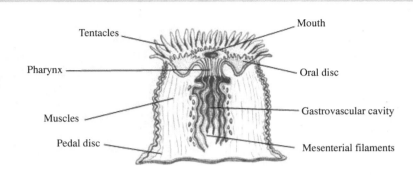

FIGURE 7.I.9
Cross-section through *Metridium*, a common sea anemone, with major morphologic features labeled.

Source: Rachel E. G. Kalisperis

Cnidarian Taxonomy

PHYLUM CNIDARIA

CLASS HYDROZOA: Both polyp and medusa in life cycle.

ORDER HYDROIDEA: hydroids, "fire coral". Polyp dominant. [*Hydra, Obelia*]

ORDER SIPHONOPHORA: free swimming or floating colonies. [*Physalia, Porpita*]

CLASS ANTHOZOA: sea anemones, corals, sea fans, soft corals.

SUBCLASS OCTOCORALLIA: soft corals.

ORDER GORGONACEA: Colonial, internal skeleton, pinnate tentacles on polyp. [*Leptogorgia*]

ORDER PENNATULACEA: sea pansies, sea pens. [*Renilla*]

SUBCLASS ZOANTHARIA: medusoid stage absent.

ORDER ACTINARIA: stony corals. External skeleton, colonial or solitary.

ORDER MADREPORARIA: sea anemones [*Metridium, Calliactis*]

CLASS CUBOZOA: sea wasps, box jellyfish.

CLASS SCYPHOZOA: true jellyfish. Medusae only—polyp generation reduced or absent. [*Aurelia*]

Ctenophores (Phylum Ctenophora)

Animals in the Phylum **Ctenophora** are often misidentified as cnidarians. The animals in this small phylum are exclusively marine, and are commonly known as sea grapes, sea gooseberries or comb jellies. While these organisms have a jelly-like, radially symmetric body with a gastrovascular cavity, there are some important differences compared to cnidarians. The phylum is named for the presence of eight equally spaced rows of ciliary comb plates, or **ctenes** (Figure 7.I.10). Each row con-

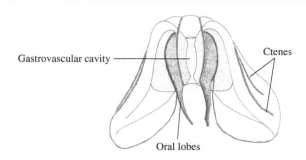

Gastrovascular cavity

Ctenes

Oral lobes

Source: Rachel E. G. Kalisperis

FIGURE 7.I.10
A ctenophore, *Beroe*, the comb jelly. Ctenes are typically iridescent and always occur in 8 rows.

sists of a succession of tiny bands formed of large cilia fused together at the ends like the teeth of a comb. The rows radiate over the surface of the animal from the upper pole to the lower pole, similar to the lines of longitude on a globe. These ctenes are iridescent and are easily seen even in the smaller comb jellies.

Flatworms (Phylum Platyhelminthes)

Members of this phylum are unsegmented, **dorsoventrally** flattened worms (hence the name flatworm), and due to the presence of the third tissue layer (the mesoderm) they display an acoelomate body structure (Figure 7.I.11). Cephalization, a trend towards a centralized nervous system, and bilateral symmetry identify this phylum as having developed major evolutionary advancements relative to the previous phyla. In most flatworms, there is no body cavity other than a digestive tract. The area between the digestive tract and the body wall is filled with **mesenchyme**, a mass of loosely packed cells. The mouth (when present) is the only opening to the digestive tract. Animals in this phylum possess no anus, and thus these organisms have an incomplete digestive tract. Flatworms are **monoecious**, meaning each worm possesses both male and female reproductive organs.

Of the three classes that comprise this phylum (refer to Figure 7.I.11), two are completely parasitic. The free-living flatworms, the turbellarians, are usually ciliated to help themselves move through aquatic habitats. Many have regenerative powers, as is seen by cutting a planarian in half; within a week, the worm will regenerate the removed portions of its body, resulting in two new worms. The trematode flukes can be either ectoparasitic (external) or endoparasitic (internal), and all possess a digestive tract. The digestive tract gives rise to one or two intestinal **caeca**, which are rounded, branched tubes in the worm's body. Suckers or other adhesive organs are often present. Many of the commonly known and medically significant flukes are parasitic in mammals, and can be transmitted to mammals through secondary hosts such as fish, birds, and frogs. The cestodes, or tapeworms, are endoparasitic in vertebrates, with no digestive tract or mouth; all nutrition is taken into the body across the epithelial layer. Most tapeworms have a **scolex**, an anterior adhesive organ with suckers and spines used to attach to the inner lining of the host's stomach. Parasitism by tapeworms can be detected by looking for

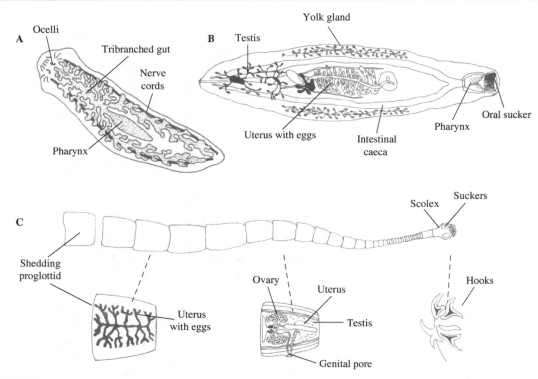

FIGURE 7.I.11

Representative flatworms. (A) A common turbellarian, the planarian *Dugesia*. (B) The trematode liver fluke *Fasciola hepatica*. (C) *Taenia*, the common tapeworm. Different proglottids have specialized functions, and are often in different reproductive stages (see first and second insets). The hooks on the scolex are shown in the far right inset.

Source: Rachel E. G. Kalisperis

proglottids, specialized segments comprising the majority of the worm's body and containing reproductive organs. These proglottids, which look like tiny grains of rice, are shed in the animal's feces.

Flatworm Taxonomy

PHYLUM PLATYHELMINTHES

CLASS TURBELLARIA: free-living flatworms. [*Bdelloura, Dugesia*]

CLASS TREMATODA: the flukes, parasitic, with a digestive tract. [*Clonorchis*]

CLASS CESTODA: the tapeworms, endoparasitic, with no digestive tract. [*Taenia*]

Phylum Nemertea (Nemertina, Rhynchocoela)

Nemerteans are commonly called ribbon worms. These animals have unsegmented, dorsoventrally flattened bodies with a unique and characteristic **proboscis** structure containing a cavity called a **rhynchocoel** (Figure 7.I.12A). The proboscis can be armed with spines, and the worms actively feed on small invertebrates by extending and coiling the proboscis around the prey. Ribbon worms are predominantly ben-

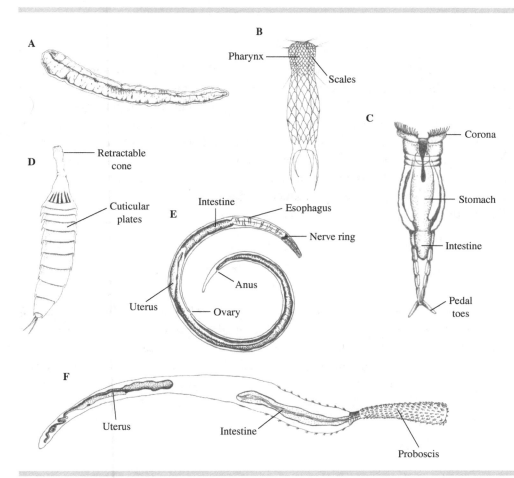

FIGURE 7.I.12
Representative phyla of the lower animals. (A) Nemertean without rhynchocoel extended. (B) A typical gastrotrich. (C) The rotifer *Philodina*. (D) *Echinoderes*, a kinorhynch. (E) The common nematode *Ascaris*. (F) An acanthocephalan. Note that illustrations are not drawn to scale.

Source: Rachel E. G. Kalisperis

thic, marine organisms, and their length ranges from centimeters to several meters. Nemerteans have developed a **true circulatory system**, in which the blood is enclosed in vessels and does not directly contact the organs or tissues. They have also developed a complete digestive tract. Nemerteans are believed to be acoelomate, though some scientists classify them as coelomate based on the presence of the rhynchocoel cavity in the proboscis and the circulatory system. Most ribbon worms are **gonochoristic**, having two separate sexes, and during mating, scores of worms will form large writhing knots, allowing them to coordinate the release of their eggs and sperm.

Phylum Gastrotricha

Gastrotrichs are small, less than 1 mm long free-living marine and freshwater organisms (Figure 7.I.12B). The body is bottle-shaped and divided into a head and trunk, with characteristic spines or scales. These organisms are found in the interstitial spaces of bottom sediments or on the surface of submerged plants and animals. They have a complete digestive tract and are classified as acoelomate. Most are **hermaphroditic**, possessing both male and female organs. The existence of males is unknown in certain groups.

Phylum Rotifera

Most rotifers are tiny, often less than 1 mm long, but are quite complex and display many different forms. Rotifers are most commonly found in freshwater. The body contains a head, trunk, and foot with "toes" that contain adhesive glands (Figure 7.I.12C). The anterior end bears a ciliated organ called the **corona**, which spins like a wheel as the animal feeds. The beating of the cilia pulls food particles into the mouth and also functions in locomotion. Members of this phylum are pseudocoelomate, with a complete gut. Reproduction is poorly known; males are reduced in size or completely absent in some classes. **Parthenogenesis** is common, where females reproduce asexually by producing eggs that develop into females without fertilization.

Phylum Kinorhyncha

The tiny kinorhynchs, most less than 1 mm in length, are found living in mud or the interstitial spaces between sand grains of marine sediments. These organisms have a characteristic segmented body composed of thick cuticular plates, with spines extending from each segment (Figure 7.I.12D). The head has a retractable cone and a ring of spines around the mouth. These organisms molt their cuticle in a manner similar to crustaceans. Kinorhynchs are gonochoristic and pseudocoelomate, with true segmentation and a complete gut.

Phylum Nematoda

Perhaps one of the most well-studied of the lower invertebrates are the nematodes. There are many parasitic forms, and this phylum is considered one of the most abundant animal groups on Earth. It has been said that if all the structures on Earth were removed, including trees, soil, water, buildings, etc., the outline of the earth and its features would still be seen because of the vast number of nematodes that inhabit every environment and surface available. Nematodes parasitize nearly every group of animal and plant. Some even inhabit human intestines, causing diseases such as elephantiasis. The more familiar parasitic nematodes include the hookworms, pinworms, and whipworms. Nematodes are unsegmented pseudocoelomates, with a round, cuticle-covered body and complete digestive tract (Figure 7.I.12E). There is some cephalization, and the anterior end contains a mouth with "lips". The body wall has only longitudinal muscles. Because of this, the nematode is unable to creep and must move by whipping its body back and forth. Most nematodes are gonochoristic, and the parasitic nematodes have a complex life cycle involving multiple larval and cyst stages.

Phylum Acanthocephala

The acanthocephalans, whose name means "spiny head", are common parasites of fishes. The digestive tract in this phylum has been completely lost, and the anterior end of the worm has developed an eversible proboscis with recurved hooks to attach to its host (Figure 7.I.12F). While most acanthocephalans are less than 20 cm long, some can achieve a meter in length. These unsegmented worms have a pseudoceolom, a complex life cycle, and separate sexes.

EXERCISE 7.1
SURVEY OF MARINE ORGANISMS

PART I

Materials required: Labeled specimens (plastomount, preserved, or live), computer, reference texts, identification or field guides, and pictures as needed

Directions

1. Representative specimens from the phyla discussed on the previous pages have been set out in stations in order of increasing complexity. Each station has a particular set of questions associated with it. Answer the questions pertaining to each organism.

2. You do not have to start with Station #1, but be sure to answer each question.

Appendix A at the end of this lab manual provides a brief introduction to the use of microscopes if you need a refresher on how to use them.

Phylum Porifera

1. What is the common name for this phylum?_____

2. a. What is the name of the openings through which water enters the organism?

    _____

   b. What is the name of the opening where water leaves the organism?

    _____

3. What is the function of the choanocytes (collar cells)? _____

    _____

4. What would be the purpose of having an increase in folding and greater numbers of canals and chambers? _____

    _____

Phylum Cnidaria

5. What characteristic is unique to this phylum? _____

   _____

6. What are the two functions for the tentacles of a hydra? _____

   _____

7. Identify the following cnidarians as either a single animal or a colonial organism:

   a. *Physalia* _____

   b. *Aurelia* _____

   c. *Leptogorgia* _____

   d. *Acropora* _____

8. What would be the purpose or benefit to being a colonial organism? _____

   _____

Phylum Ctenophora

9. List two purposes or functions of the comb rows on a ctenophore: _____

   _____

Phylum Platyhelminthes

10. **Planaria:** Compared to the "simpler" phyla, what special feature do these worms possess?

    _____

Phylum Rotifera

11. a. How do rotifers obtain their food? _____

    _____

    b. Many rotifers can extend many times their body length. What would be the purpose of this adap-
    tation? _____

    _____

Phylum Kinorhyncha

12. What two features of the Kinorhyncha help it to move and live interstitially between sand grains on the beach? _____

_____

Phylum Gastrotricha

13. Gastrotrichs are also interstitial organisms in beach sand. What are two characteristics that distinguish it as a separate phylum from the kinorhynchs? _____

_____

_____

Phylum Nematoda

14. Nematodes are important and common parasites found in nearly every group of animal. What particular feature of a nematode can be used to distinguish it from any other kind of worm?

_____

_____

Additional Information

If you are interested in further information about the topics in this lab, the following sources will provide you with detailed descriptions, identifying characteristics, and interesting features of organisms in the invertebrate animal phyla:

- ▶ Barnes, R. D. 1974. *Invertebrate Zoology.* 3rd ed. Philadelphia: Saunders Publishers.
- ▶ Brusca, R. C., Moore, W. and Shuster, S. M. 2016. *Invertebrates.* 3rd ed. Sunderland, MA: Sinauer Associates.
- ▶ Maddison, D. R. 2001. The Tree of Life Web Project. http://www.tolweb.org.

LAB 7

SURVEY OF MARINE ORGANISMS
PART II

OBJECTIVES

▶ Learn the identifying characteristics of the major marine phyla of invertebrate organisms and be able to recognize these groups and some of their subgroups.

▶ Understand the differences between protostome and deuterostome development.

▶ Be able to identify general evolutionary relationships and trends seen in these invertebrate groups and how these apply to the marine environment.

POINTS TO PONDER

▶ Why are only eucoelomates categorized into protostomes and deuterostomes?

▶ Why do some gastropods have such ornate and brightly colored shells?

▶ What characteristics of arthropods allow them to be the most numerous organisms on Earth?

TERMS TO KNOW

aboral	Bivalvia	cleavage pattern	Echinodermata
ambulacral groove	blastopore	Copepoda	Echinoidea
Amphipoda	Bryozoa	Crustacea	exoskeleton
Annelida	Cephalochordata	cuticle	foot
antennae	Cephalopoda	determinate	ganglia
antennule	Chaetognatha	development	Gastropoda
Aristotle's lantern	chitin	deuterostome	gonochoristic
Arthropoda	Chordata	dorsal hollow nerve	Hemichordata
Asteroidea	Cirripedia	cord	hemolymph

Hirudinea	metamerism	Polychaeta	tunicate
Holothuroidea	Mollusca	Polyplacophora	umbo
incurrent siphon	molting	protostome	Urochordata
indeterminate	nephridia	radial cleavage	veliger
development	notochord	radula	Vertebrata
Isopoda	Oligochaeta	setae	visceral mass
lophophore	Ophiuroidea	shell	water vascular
madreporite	ossicle	spiral cleavage	system
Malacostraca	Ostracoda	torsion	zooid
mandibles	parapodia	triploblastic	
mantle	pharyngeal clefts	trochophore	
mantle cavity	polar axis	tube foot	

SURVEY OF THE INVERTEBRATE ANIMAL PHYLA, PART II

In Part I of this lab, the survey of marine organisms introduced you to many of the "lower" animal phyla such as the sponges and cnidarians. You were also introduced to the **triploblastic** organisms, those with three tissue layers that were categorized based on a particular body cavity type: acoelomate, pseudocoelomate, or eucoelomate. The eucoelomate phyla will be examined in this second part of the invertebrate phyla survey.

The organisms that have a true coelom, the eucoelomates, may be divided into two groups, the **protostomes** and the **deuterostomes**. Each group has a characteristic mode of development. In the protostomes, the first opening that develops (the **blastopore**) leading into the digestive cavity becomes the mouth ("proto" = first, "stome" = mouth). Annelids, mollusks, and arthropods are major protostome groups. In the deuterostomes ("deutero" = other), the blastopore becomes the anus, and the mouth forms as a second opening some distance away. Echinoderms, hemichordates, and chordates are deuterostome groups. While some of the lower invertebrate phyla have developmental characteristics similar to the groups, by definition *only* eucoelomate organisms can be classified as protostomes and deuterostomes.

Protostomes and deuterostomes also differ in the pattern in which the fertilized egg divides into smaller cells, called the **cleavage pattern** (Figure 7.II.1). Protostomes undergo **spiral cleavage**, in which the cell cleavage planes are tilted to the long axis (**polar axis**) of the egg, resulting in a spiral cell arrangement. The fate of each cell in a protostome embryo is irreversibly determined early in development. This is called **determinate development**. Deuterostomes show **radial cleavage** where the cleavage plane is perpendicular to the long axis of the egg. If the cells of a deuterostome embryo are separated, each cell can form a complete individual; this is called **indeterminate development**. The same experiment performed on a protostome embryo will result in the cell producing a predetermined quarter of an individual—each cell grows into the portion of the organism it would have formed if left together.

	8 cell stage, side view	8 cell stage, polar view
Deuterostome radial cleavage		
Protostome spiral cleavage		

FIGURE 7.II.1

Comparison between deuterostome radial cleavage and protostome spiral cleavage developmental patterns, as viewed from the side and top (polar axis). Arrows denotes direction of cleavage for an individual cell.

Source: Michelle L. Hardee

THE PROTOSTOMES

Annelids (Phylum Annelida)

Members of this phylum exhibit **metamerism,** or true segmentation, an important developmental adaptation. The body is divided into a series of segments, all similar structurally, with repeating organs and organ systems in each segment. The development of metamerism is important because it allows for specialization of body regions; groups of segments can become modified for different functions as seen in higher phyla. In addition, compartmentalization of the body cavity allows the use of the coelom as a hydrostatic skeleton, in which the coelomic fluid is used to provide support (instead of, for example, the exoskeleton of a crab). The annelids possess many other important characteristics, including a complete digestive tract, a closed circulatory system with oxygen-carrying hemoglobin, and specialized excretory tubules called **nephridia**. Their nervous system is centralized with a pair of brain-like cerebral **ganglia,** and segmental ganglia appear in each segment. The size of annelids can range from less than 1 mm to 3 m (the giant Australian earthworm).

There are three classes of annelids: **Polychaeta, Oligochaeta,** and **Hirudinea** (Figure 7.II.2). Members of the Polychaeta ("many setae") are quite diverse and almost all are marine. Polychaete worms possess paddle-like appendages termed **parapodia,** which contain hair-like bristles called **setae** (see inset in Figure 7.II.2A). These parapodia can function either as gills or for locomotion. Most polychaetes reproduce sexually, and the majority are **gonochoristic** (separate sexes). The oligochaetes ("few setae") are almost all freshwater or terrestrial. They possess short setae but lack parapodia for movement, instead moving by muscle contraction. Oligochaetes are always monoecious. The majority, including the common earthworm, *Lumbricus*, are scavengers that feed on dead organic matter.

The hirudineans are more commonly known as leeches and lack both setae and parapodia. At the anterior and posterior ends of these animals are suckers that serve for attachment. The mouth is inside the anterior sucker, while the anus is outside the posterior sucker. Many leeches are predatory, but approximately three-quarters are blood-sucking ectoparasites. Predatory leeches feed on small invertebrates such

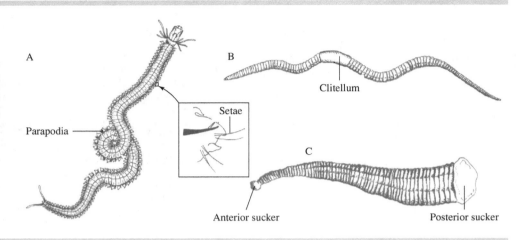

FIGURE 7.II.2
The major annelid classes. (A) A common marine polychaete, *Nereis*. A parapodium has been expanded to show the setae. (B) The earthworm, *Lumbricus*, an oligochaete. (C) A typical hirudinean, the leech.

Source: Rachel E. G. Kalisperis

as snails and insect larvae, swallowing their prey whole. The parasitic species feed on numerous hosts, including both invertebrates and vertebrates. These leeches secrete both an anesthetic and an anticoagulant (hirudin) when attached to the host, to prevent the blood from clotting.

Annelid Taxonomy

PHYLUM ANNELIDA

> CLASS POLYCHAETA: polychaetes worm. Parapodia present, segments with long setae. [*Nereis*, *Arenicola*]

> CLASS OLIGOCHAETA: earthworms. Parapodia absent, segments with short setae. [*Lumbricus*]

> CLASS HIRUDINEA: leeches. Parapodia and setae absent, anterior and posterior suckers present.

Mollusks (Phylum Mollusca)

A walk along any shoreline provides the casual beachcomber with numerous examples of the Phylum Mollusca. Mollusks are the second largest invertebrate phylum after the arthropods; over 80,000 living species of mollusks have been described. The presence of a shell has led the mollusks to be well established in the fossil record as well; some 35,000 fossil species have been identified. Representatives from the molluscan classes are diverse, yet all are built on the same fundamental body plan. The generalized body plan of mollusks has evolved into characteristic morphologies that allow these organisms be classified into six major molluscan classes. Out of these, the four most common ones will be examined here.

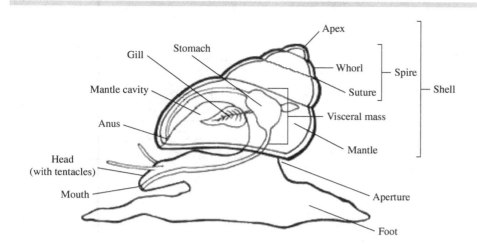

FIGURE 7.II.3
A generalized mollusk with characteristic internal and external morphologic features labeled.

Source: Rachel E. G. Kalisperis

All mollusks are characterized by a hard shell, mantle, foot, and visceral mass (Figure 7.II.3). The generalized mollusk (above) is bilaterally symmetrical, with a well-defined head bearing a pair of tentacles. The ventral surface of the body forms a muscular **foot** used for locomotion. The **visceral mass** contains most of the internal organs, and a layer of tissue, called the **mantle**, drapes over the visceral mass. The mantle also extends beyond the visceral mass, creating a water-filled chamber called the **mantle cavity** that houses the gills, anus, and specialized excretory nephridia. The dorsal surface is covered by a protective **shell**, which is secreted by the mantle. Mollusks lack the true segmentation found in annelids. Many mollusks feed with a strap-like rasping organ called a **radula**, which contains rows of recurved chitinous teeth used to scrape up algae, detritus, and other food particles. Most marine mollusks have a ciliated free-swimming **trochophore** larva. In many molluscan classes, the trochophore passes into a more highly developed **veliger** larval stage in which the foot, shell, and other structures appear.

1. Class Polyplacophora

This class includes the chitons that are common along rocky shorelines and in tidal pools. Chitons have a large foot and a shell composed of eight overlapping dorsal plates (the body itself is not segmented) (Figure 7.II.4A). Their foot and shell provide them with protection in their habitat, providing a shield and suction to prevent them from being dislodged by crashing waves.

2. Class Bivalvia

The bivalves (also called Class Pelecypoda) includes mollusks with two equal shells that are hinged together at the **umbo** (Figure 7.II.4B). The hatchet-shaped foot

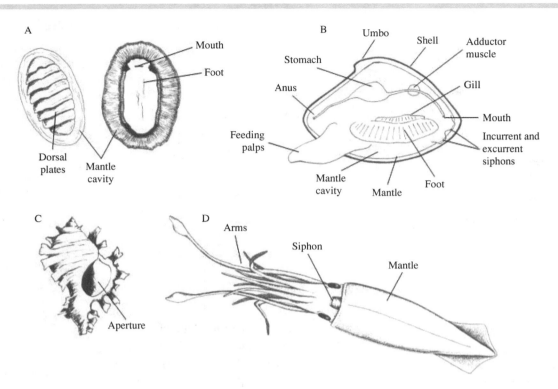

FIGURE 7.II.4
Some major Molluscan classes. (A) A representative polyplacophoran, the chiton. Both dorsal and ventral views are shown. (B) General diagram of a clam, member of Class Bivalvia. (C) *Murex,* the oyster drill, a gastropod. (D) The common squid, *Loligo,* a cephalopod. Illustrations are not to scale.

Source: Rachel E. G. Kalisperis

can be used for locomotion in sediments. Bivalves have no distinct head, and the radula has been lost. To feed, bivalves pull water through an **incurrent siphon** and across the cilia-covered gills, filtering particles. Bivalves exhibit a wide range of sizes, from the huge, meter-wide giant clams to the tiny centimeter-long *Coquina* clams found along the beach. The common oyster (*Crassostrea*) is a bivalve, forming oyster reefs by clumping together in groups and reducing the size of one shell. As oysters feed, they filter out detritus and pollutants from the seawater. A single oyster can clean ~50 gallons of water per day!

3. Class Gastropoda

Snails belong to the largest mollusk class, the gastropods ("stomach-foot"), characterized by a single spiral shell. A distinctive feature of the gastropods is a developmental process called **torsion**. During embryonic development, one side of the visceral mass grows faster than the other. This uneven growth causes the body organs to rotate 180°, placing the anus above the head in the adult (see Figure 7.II.3). Gastropod shells often have intricate patterns, colorations, ribs, or spines, such as *Murex,* the oyster drill (Figure 7.II.4C). Ornate spines are believed to function as support to prevent the shell from sinking in soft-bottomed habitats, and also to deter predators. Some marine gastropods such as the horse conchs may reach nearly two feet in length, and others are as small as a sand grain (the pelagic pteropods). Cone shells have a dart-shaped radula that secretes a neurotoxin potent enough to kill a human.

4. Class Cephalopoda

Members of this molluscan class include the squid, cuttlefish, and the octopus. Cephalopods (meaning "head-foot") are typically characterized by having an elongated body, with 8 to 10 arms surrounding the mouth (Figure 7.II.4D). Most cephalopods have either lost their shell during evolution (the octopus), or it has been reduced and internalized into a "pen" (squid and cuttlefish). The oval-shaped pen of the cuttlefish is often given to pet birds as a calcium supplement. The radula has been modified into a sharp beak, and some octopus species use it to inject poison into their prey. Cephalopods have a unique method of rapid locomotion, called jet propulsion, in which water is sucked into the mantle cavity and blown out forcefully to propel the animal forward. Cephalopods range in size from tiny octopuses to the largest of all invertebrates, the giant squid. Their size can be upwards of 17 m (including the tentacles), weighing 2 tons. Octopuses have a complex brain, and are able to solve simple puzzles and mazes, as well as open jars and manipulate objects.

Mollusk Taxonomy

Phylum Mollusca

Class Monoplacophora: [*Neopilina*]

Class Polyplacophora: chitons. Shell with 8 dorsal plates.

Class Gastropoda: snails, slugs, limpets, nudibranchs, sea hares. One shell, or shell secondarily absent, occurrence of torsion during development. [*Littorina, Busycon, Aplysia*]

Class Bivalvia (Pelecypoda): mussels, clams, oysters, scallops. Two shells, hinged, no radula. [*Crassostrea, Mercenaria, Tagelus*]

Class Scaphopoda: tusk or tooth shells. Conical shell open at both ends.

Class Cephalopoda: squid, octopi, chambered *Nautilus*, cuttlefish. Shell lost or small and internal. [*Loligo*]

PHYLUM ARTHROPODA—THE CRUSTACEANS

The Phylum Arthropoda is a vast assemblage of animals, with nearly one million described species. They are present in nearly all habitats on Earth and are the most successful phylum of animals to ever live. Arthropods represent the evolutionary culmination of the protostomes. They arose either from a primitive annelid or from an ancestor common to both. True segmentation (metamerism) is evident in the embryonic development of all arthropods and is a conspicuous feature of many adults. However, within a number of the arthropod subphyla (arachnids and crustaceans) there has been a tendency for metamerism to disappear. In addition, appendages on different segments have become structurally and functionally differentiated.

The tremendous adaptive diversity and success of arthropods is a result of the characteristic features of the phylum that enables arthropods to survive in virtually every habitat. All arthropods possess: (1) a hard, chitinous exoskeleton and (2) jointed appendages, along with some level of segmentation. The body of an arthropod is completely covered by the **cuticle,** which is an external skeleton or **exoskeleton** constructed out of a type of carbohydrate called **chitin.** The cuticle can be thick and hard in some regions, and soft and pliable in others, such as the joints. The exoskeleton is not only protective but also provides attachment points for the muscles. However, the rigidity of the exoskeleton becomes a problem during growth. Arthropods must occasionally shed the old exoskeleton and secrete a larger one. This **molting** process is energetically expensive and leaves the animal temporarily vulnerable to predation. Groups of segments and their associated jointed appendages have become specialized for a variety of functions. During evolution, such specialization was key to the diversification of arthropods as well as allowing the division of labor among body regions. For example, there are different appendages for walking, feeding, sensory reception, defense, and even reproduction (see Figure 7.II.5).

Arthropods have a specialized nervous system and sensory organs with a high degree of cephalization. Their digestive, reproductive, and excretory systems are well developed and often specialized for living in water and overcoming the limitations of the exoskeleton. Arthropods have an open circulatory system, in which the **hemolymph** fluid is pushed by a heart through vessels and then into sinus spaces surrounding the tissues and organs. The hemolymph then returns to the heart through pores.

The arthropods include many different categories of animals, such as arachnids (spiders), insects, and centipedes. Marine arthropods, or **crustaceans,** are probably one of the most recognized invertebrate groups, because this group includes the shrimp, crabs, and lobster often found on seafood lovers' plates (Figure 7.II.5). There are more than 30,000 described living species of crustaceans, with many more waiting to be discovered. Crustaceans exhibit an incredible diversity in shape, form, and function, ranging from tiny planktonic larvae to giant crabs (with leg spans of up to 4 m). Such a range of diversity exceeds even that of the insects. Despite such diversity, all crustaceans still display the characteristic features unique to their phylum, and since the Crustacea are the major arthropods in the marine environment, we shall deal exclusively with them.

The crustaceans are the aquatic branch of the mandibulate arthropods, an evolutionary lineage that includes the uniramians (insects and centipedes). These arthropods have developed a pair of specialized jaw-like chewing appendages called **mandibles.** The other main group of arthropods, the chelicerates (spiders, scorpions, and horseshoe crabs), have pedipalps and chelicerae as their sensory and feeding appendages, respectively, instead of mandibles. Mandibulate arthropods are further distinguished from chelicerates by having either one pair (insects) or two pairs (crustaceans) of sensory **antennae.** The second pair of antennae in the crustaceans is called the 2nd

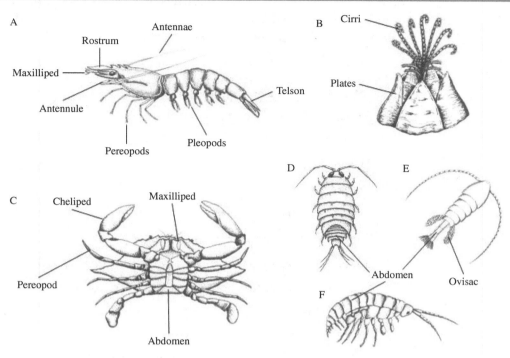

FIGURE 7.II.5

Arthropod appendages are typically modified for different functions as exemplified in these representative crustaceans. (A) A common penaeid shrimp. (B) *Balanus*, an acorn barnacle. The cirri are modified hind legs for use in feeding. (C) *Callinectes*, the blue crab, ventral view. Note the reduced and modified appendages and abdomen when compared to the shrimp. (D) An isopod. (E) *Cyclops*, a copepod. (F) *Gammarus*, an amphipod. The antennae are elongated and the abdomen reduced in these smaller crustaceans, which primarily live in the open ocean. Illustrations are not to scale.

Source: Rachel E. G. Kalisperis

antennae or **antennules**. The absence of antennules in the insects and multi-leg forms is an important identifying feature separating them from the crustaceans.

To do justice to the major crustacean groups would entail descriptions that fill up many more pages than can be provided here. What follows is a much-abbreviated list of arthropod taxa with a few general characteristics. Not all the taxa on this list will be examined—it is for your reference and to gain an appreciation of the diversity present in the Crustacea. You should, however, be familiar with some of the more common groups of crustaceans: the ostracods (**Ostracoda**), copepods (**Copepoda**), barnacles (**Cirripedia**), isopods (**Isopoda**), amphipods (**Amphipoda**), and the familiar members of Class **Malacostraca** (shrimp, crabs, and lobsters).

Arthropod Taxonomy

PHYLUM ARTHROPODA

SUBPHYLUM CHELICERIFORMES: spiders, scorpions, mites, ticks, horseshoe crabs, sea spiders. Body divided into two regions, the cephalothorax and abdomen, no antennae, with 6 pair of appendages: chelicerae, pedipalps, and 4 pair of legs.

SUBPHYLUM UNIRAMIA: insects (flies, roaches, beetles, bees, fleas, etc.), millipedes and centipedes. Abdomen with many segments, head with 1 pair antennae, appendages not branched (uniramous).

SUBPHYLUM CRUSTACEA: crabs, shrimp, lobsters, copepods, barnacles. Head (cephalon) with 5 pair appendages: 1st and 2nd antennae, mandibles, 1st and 2nd maxillae, trunk (thorax and abdomen) with multiple pairs of appendages (pereopods and pleopods), nauplius larvae.

CLASS BRANCHIOPODA:

ORDER ANOSTRACA: fairy or brine shrimp. [*Eubranchiopus, Artemia*]

ORDER CLADOCERA: water fleas. [*Daphnia*]

CLASS MAXILLOPODA:

SUBCLASS OSTRACODA: ostracods. [*Cypris*]

SUBCLASS COPEPODA: copepods. [*Calanus, Cyclops, Harpacticus*]

SUBCLASS CIRRIPEDIA: gooseneck and acorn barnacles. [*Lepas, Balanus*]

CLASS MALACOSTRACA:

SUBCLASS EUMALACOSTRACA: shrimp, crabs, lobsters.

ORDER STOMATOPODA: mantis shrimp. [*Squilla*]

ORDER EUPHAUSIACEA: krill. [*Euphausia*]

ORDER DECAPODA: crabs, shrimps, lobsters

SUBORDER DENDROBRANCHIATA (NATANTIA): penaeid (commercial) shrimp. Do not brood eggs in pouches. [*Penaeus, Cranenon*]

SUBORDER PLEOCYEMATA: Brood eggs in pouch under abdomen.

INFRAORDER ASTACIDEA: crayfish, large-clawed lobsters. [*Homarus*]

INFRAORDER PALINURA: spiny lobsters. [*Panulirus*]

INFRAORDER ANOMURA: hermit crabs, porcelain crabs, mole crabs. Abdomen is soft or short and slightly flexed beneath the thorax. [*Emerita, Pagurus, Clibanarius*]

INFRAORDER BRACHYURA: true crabs. Abdomen reduced and flexed beneath thorax. [*Callinectes, Menippe, Cancer*]

SUPERORDER PERACARIDA: pouched or brood shrimps

ORDER ISOPODA: isopods (rock lice, pill bugs, etc.). Dorsoventrally flattened. [*Bathynomas, Ligia*]

ORDER AMPHIPODA: amphipods. Laterally flattened. [*Gammarus*]

THE DEUTEROSTOMES

Phylum Echinodermata

Echinoderms are exclusively marine and are relatively large benthic invertebrates, the smallest being about a half-inch in diameter. The most notable characteristic of this group is their pentamerous radial symmetry, in which the body is divided into five parts arranged around a central disc. This radial symmetry, however, has been secondarily derived from an ancestral *bilateral* form; the echinoderms are in no way related to the other radially symmetric phyla (cnidarians and ctenophores). Furthermore, echinoderms are true coelomate animals and have a higher level of complexity than the other radiate groups. Most echinoderms exhibit amazing regenerative powers and can regrow arms and other body parts that have been cast off or injured when seized by a predator.

All echinoderms have a characteristic internal skeleton (endoskeleton). This skeleton is composed of calcareous **ossicles** or plates which typically have projecting spines— hence the name echinoderm, or "spiny skin". Another unique feature of echinoderms is the presence of a **water vascular system.** This is a network of water-filled internal canals and pores that connects to the outside through a button-shaped **madreporite** located on the **aboral** surface, the side opposite the mouth (Figure 7.II.6A). Some canals terminate in a bulbous **tube foot.** The tube feet are arranged in rows along the

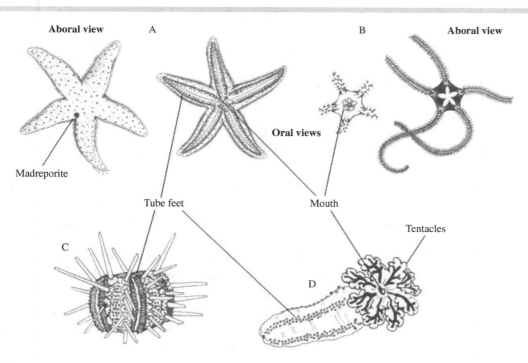

FIGURE 7.II.6

Echinoderms. (A) Generalized diagram of Class Asteroidea, the sea star, with aboral and oral view. (B) An ophiuroid brittle star, with oral and aboral view. (C) An echinoid, the sea urchin. (D) The sea cucumber, a holothuroid echinoderm.

Source: Rachel E. G. Kalisperis

length of the **ambulacral groove** and act as suckers to aid the animal in locomotion and food gathering.

1. Class Asteroidea

The Asteroidea are the echinoderms commonly known as starfish or sea stars (Figure 7.II.6A). The body is composed of a central disc from which the arms extend. Sea stars typically possess five arms, although some, such as the sun stars, may have 40 or more. Sea stars are carnivorous and feed on mollusks, crustaceans, polychaetes, and even fish. In order to feed, the sea star uses its tube feet to pull apart the shells of a clam, and then everts its stomach into the interior of the clam. The soft parts of the prey are reduced to a liquid by external digestion, which is then passed into the sea star's body.

2. Class Ophiuroidea

The class Ophiuroidea contains the echinoderms known as brittle stars (Figure 7.II.6B). While ophiuroids resemble sea stars, the extremely long arms of ophiuroids are more sharply defined at their connection to the central disc. There is no ambulacral groove, and the tube feet play little role in locomotion. The arms appear jointed due to the presence of longitudinal rows of calcareous plates that are part of the endoskeleton. Brittle stars move by looping their arms along a substrate and pulling the body forward. To feed, brittle stars can curl their arms around food particles, and rake food into their mouth.

3. Class Echinoidea

The echinoids are free-moving echinoderms commonly known as sea urchins, heart urchins, and sand dollars (Figure 7.II.6C). The echinoid body does not have arms like sea stars. The name Echinoidea means "like a hedgehog", and is derived from the fact that the bodies of these animals are covered with spines. Sea urchins are armed with long movable spines, whereas the other echinoids that are adapted for burrowing in sand (heart urchins and sand dollars) have much smaller spines. The echinoid shape is circular and the body is either spherical or flattened (as in sand dollars) along the oral-aboral axis. The skeletal ossicles are fused into a solid endoskeleton. Sea urchins also have a highly developed chewing apparatus called **Aristotle's lantern,** which is composed of five calcareous structures. Sand dollars possess a modified lantern; you may have seen the pieces (often referred to as the five "doves") that spill out when a sand dollar is broken in half.

4. Class Holothuroidea

The holothuroids, or sea cucumbers, also lack arms, and the mouth and anus are located at opposite ends of an elongated, cucumber-shaped body (Figure 7.II.6D). The skeleton is reduced to microscopic ossicles, and the tube feet are arranged in longitudinal rows. Some of the tube feet are modified into a circle of tentacles around the mouth. Sea cucumbers are sluggish animals and either live on the seafloor surface or burrow in sand and mud. When attacked by a predator, many holothuroids display the ability to eviscerate, or disgorge, their digestive tract. This allows the holothuroid to escape while its predator is feeding on the eviscerated organs. The digestive parts are later regenerated. Sea cucumbers are unusual in that they breathe through their anus, because their respiratory organs branch from the end of the digestive tract.

Echinoderm Taxonomy

Phylum Echinodermata

Class Asteroidea: sea stars. [*Asterias*]

Class Ophiuroidea: brittle stars, basket stars. [*Ophiothrix*]

Class Echinoidea: sea urchins, sand dollars. [*Lytechinus*, *Mellita*]

Class Holothuroidea: sea cucumbers. [*Thyone*]

Phylum Bryozoa (Ectoprocta)

With few exceptions, the bryozoans ("moss animals") are colonial and sessile animals (Figure 7.II.7A), composed of many individual **zooids** linked together that serve a function similar to the polyp of cnidarians. The phylum consists of three classes with representatives in both marine and freshwater environments. Most species are found in coastal waters and form bushy or encrusting colonies attached to rocks, shells, algae, and other animals. Bryozoans are characterized by having a retractable **lophophore**, a horseshoe-shaped ciliated feeding structure with hollow tentacles.

Phylum Chaetognatha

The chaetognaths, also known as arrow worms, are common marine plankton (Figure 7.II.7B). All 65 species in this phylum are marine and planktonic. The adults possess none of the features common to other deuterostome phyla and are very similar to pseudocoelomates; only during their development do arrow worms show deuterostome features. These organisms are voracious predators of other zooplankton and possess a set of sharp, curved spines or "jaws" surrounding their mouth that quickly clamp shut around their prey, much like a Venus fly-trap.

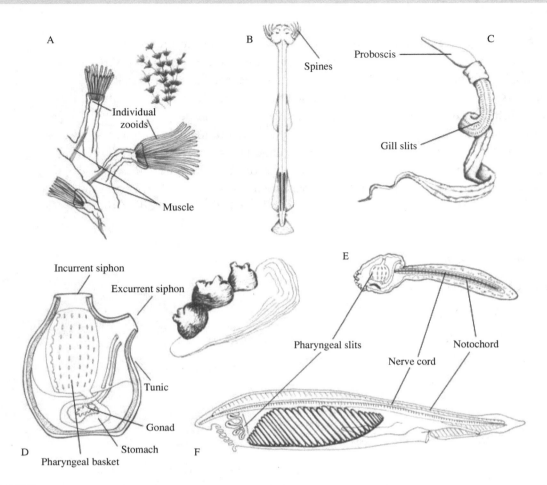

FIGURE 7.II.7

Higher deuterostomes. (A) Close-up of a typical bryozoan zooid; inset is a representation of the larger colony. (B) The carnivorous chaetognath, *Sagitta*. (C) The hemichordate *Saccoglossus*. (D) A common tunicate, *Molgula;* also shown is an oyster shell colonized by several sea squirts. (E) A general diagram of a tunicate larva. Note the characteristic chordate features. (F) The lancelet *Amphioxus*, a representative chordate. Illustrations are not to scale.

Source: Rachel E. G. Kalisperis

Phylum Hemichordata

The hemichordates are a small group of worm-like marine animals (Figure 7.II.7C) that was once considered a chordate subphylum. The relationship with the chordates was based on the presence in the hemichordates of gill slits and what was thought to be a **notochord** (see below). It is now agreed that the hemichordate "notochord" is not a true chordate notochord, and other than the presence of **pharyngeal clefts**, the groups are quite different. The hemichordates have since been removed from the Chordata and have been given the rank of a separate phylum. This group is composed of two classes: the Enteropneusta (acorn worms) and the Pterobranchia. The acorn worms are the most common and best-known hemichordates.

Phylum Chordata

The Phylum Chordata is probably the most familiar animal phylum because it includes the largest and most conspicuous animals, such as fish, snakes, humans, etc. However, these animals belong to the Subphylum Vertebrata, and the Vertebrata are only one of three subphyla of Chordata. The other two (Urochordata and Cephalochordata) are related to each other and to vertebrates by possessing three common characteristics: (1) a notochord, (2) a **dorsal hollow nerve cord**, and (3) pharyngeal clefts or gill slits. The notochord provides structural support for the body, and the pharyngeal clefts function in digestion or respiration, depending on how the structure is modified. Although the Urochordata and the Cephalochordata are not as well-known as the vertebrates, they are entirely marine and may be quite common.

1. Subphylum Urochordata

Adult urochordates, commonly known as **tunicates**, have little resemblance to other chordates. Tunicates are typically barrel-shaped animals attached by one end to the substratum (Figure 7.II.7D). Only the larval stage, called the tadpole larva, possesses distinct chordate characteristics, including part of a notochord and a dorsal hollow nerve cord in the tail region that is lost in the adult (Figure 7.II.7E). The urochordate subphylum consists of three classes; most species are the ascidians (sea squirts) and can often be found below the water line on boat docks and pilings.

2. Subphylum Cephalochordata

This subphylum contains only a few closely related forms, all commonly termed lancelets. The most familiar and common genus is *Amphioxus* (Figure 7.II.7F). They are translucent and occur in shallow marine waters. Despite the fish-like appearance, they are much more primitive, as there are no paired fins of any sort. The cartilage-like notochord material stiffens the body, but no vertebrate structures (skull, ribs, vertebrae) are present. The notochord is the main skeletal structure and persists throughout life.

3. Subphylum Vertebrata

The subphylum Vertebrata includes animals with a vertebral column consisting of bony (or cartilaginous) vertebrae enclosing a spinal cord. The notochord exists only in the embryo. There are eight major vertebrate classes, each with particular characteristics.

Chordate Taxonomy

PHYLUM CHORDATA

    SUBPHYLUM UROCHORDATA: tunicates. [*Molgula*]

    SUBPHYLUM CEPHALOCHORDATA: lancelets. [*Amphioxus*]

    SUBPHYLUM VERTEBRATA

        CLASS AGNATHA: lampreys, hagfishes. Jawless, cartilaginous skeleton.

        CLASS CHONDRICHTHYES: Jawed fishes, cartilaginous skeleton.

            SUBCLASS ELASMOBRANCHII: sharks, skates, rays

        CLASS OSTEICHTHYES: Bony, jawed fishes.

            SUBCLASS ACTINOPTERYGII: Ray-finned fishes

            SUPERORDER TELEOSTEI: Most fishes, seahorses.

        CLASS AMPHIBIA: frogs, toads, salamanders. Lungs appear, gills generally absent in adult, lay shell-less eggs in water.

        CLASS REPTILIA: snakes, alligators. Primarily terrestrial, lay shelled eggs on land.

        CLASS AVES: birds. Feathers, arms modified as wings, unique lung structure.

        CLASS MAMMALIA: kangaroos, dogs, dolphins, elephants, humans, etc. Have hair, bear live young (except a few primitive forms), nurse young.

EXERCISE 7.II
SURVEY OF MARINE ORGANISMS

PART II

Materials required: Labeled specimens (plastomount, preserved, or live), computer, reference texts, identification or field guides, and pictures as needed

Directions

1. Representative specimens from the phyla discussed on the previous pages have been set out in stations. Each station has a particular set of questions associated with it. Answer the questions pertaining to each organism.

2. You do not have to start with Station #14, but be sure to answer each question.

Phylum Annelida

15. What unique features distinguish each of these worm classes from the other annelid classes?

    a. **Polychaeta:** _____

    _____

    b. **Oligochaeta:** _____

    _____

    c. **Hirudinea:** _____

    _____

Phylum Mollusca

16. **Class Bivalvia:** The clam *Mercenaria* has two small siphons used for feeding, while the oyster *Crassostrea* has none. Why? (Hint: Consider their habitats.)

_____

_____

17. **Class Gastropoda:** Examine the different types of gastropod shells. What benefits do you think the differences in these shell types provide to their respective organisms? _____

_____

_____

18. **Class Polyplacophora:** How does the shell of a chiton differ from other mollusk shells?

_____

_____

19. **Class Cephalopoda:** How do these organisms move through the water column? (Hint: Think of their main source for locomotion.) _____

_____

_____

Phylum Arthropoda

Subphylum Crustacea

20. **Class Maxillopoda:**
    a. How do barnacles feed?_____

    _____

    b. How do these organisms, which are attached to a substrate, reproduce?_____

    _____

21. **Class Malacostraca, Order Decapoda**

    a. Examine the appendages of the brown shrimp and the mole crab. How does the shrimp move compared to a mole crab? _____

    _____

    _____

    b. Examine the tail of the lobster. Why is it useful for this animal to have a segmented abdomen? How is the abdomen used for locomotion?_____

    _____

    _____

c. Based on the appendages possessed by these organisms, describe the differences in locomotion between the two animals listed below.

Blue crab: _____

_____

Stone crab: _____

_____

d. Fiddler crab: What is the purpose for having such a large claw? How does the fiddler crab feed?

_____

Subphylum Cheliceriformes

22. Horseshoe crab: Why are these crabs not classified as "true crabs"? Explain based on their appendages and body sections. _____

_____

Phylum Echinodermata

23. Describe two differences that you would use to distinguish between the two organisms listed below:

Sea star: _____

_____

Brittle star: _____

_____

24. Describe the feeding strategies used by each of the organisms listed below:

Sea urchin: _____

_____

Sand dollar: _____

_____

25. Sea cucumber: What identifiable features does this animal possess in some fashion that would help you to classify it as an echinoderm? _____

_____

Phylum Bryozoa

26. Is this a colonial organism, or are all the zooids separate animals? What is the benefit for bryozoans to be this way?

_____

Phylum Chordata

27. Tunicate or sea squirt: When a sea squirt "squirts," what structure produces the water stream?

_____

_____

Additional Information

If you are interested in further information about the topics in this lab, the following sources will provide you with detailed descriptions, identifying characteristics, and interesting features of organisms in the invertebrate animal phyla:

- ▸ Barnes, R. D. 1974. *Invertebrate Zoology*. 3rd ed. Philadelphia: Saunders Publishers.
- ▸ Brusca, R. C., Moore, W. and Shuster, S. M. 2016. *Invertebrates*. 3rd ed. Sunderland, MA: Sinauer Associates.
- ▸ Maddison, D. R. 2001. The Tree of Life Web Project. http://www.tolweb.org.

LAB 8

ADAPTATIONS OF MARINE ORGANISMS

OBJECTIVES

- ▶ Compare and contrast general adaptations between marine invertebrates and fishes.
- ▶ Learn to predict lifestyle, habitat, and other organism characteristics based on its body form.
- ▶ Understand the benefits that different adaptations provide the organisms that possess them.

POINTS TO PONDER

- ▶ Why are the fastest open ocean fishes always tapered and torpedo-shaped?
- ▶ Why are there so many different shapes of fishes?
- ▶ Why don't invertebrates have swim bladders like the fishes?

TERMS TO KNOW

anguilliform	dorsal	inferior	solute
benthic	epifaunal	lateral line	subterminal
bioluminescent	filiform	nekton	superior
carnivore	filter-feeding	neuston	suspension-feeding
caudal	fusiform	operculum	swim bladder
chromatophore	gill arches	osmoconformer	taeniform
compressiform	globiform	osmoregulator	terminal
countershading	herbivore	osmotic balance	ventral
deposit-feeding	heterocercal	pelagic	vertical migration
depressiform	homocercal	plankton	
desiccation	ichthyology	sagittiform	
detritivore	infaunal	sessile	

ADAPTATIONS OF MARINE INVERTEBRATES

As evidenced from fossils, the first marine invertebrate arose during the Cambrian period approximately 570 million years ago. As a result, marine invertebrates have had eons of time to evolve adaptations for living in the marine environment. Since invertebrates were examined in earlier labs, we will look at only a few invertebrate adaptations here. The following is an abbreviated list of features and characteristics to look for and think about when examining the marine invertebrate specimens provided in this laboratory.

1. INVERTEBRATE MORPHOLOGY

The body shape of an organism is a direct reflection of its habitat and *lifestyle*. Even though the vast ocean appears to provide relatively few habitats to which organisms can adapt, such appearances are misleading, as there is a tremendous range in the types of lifestyles found in marine invertebrate groups:

- ▶ **Pelagic**: Pelagic organisms are those that exist in the water column, away from the seafloor. They tend to be suited to moving quickly through the water or for maintaining their buoyancy in the water column. For example, squid are suited for the open ocean habitat by being streamlined and torpedo (fusiform) shaped, while many of the zooplankton such as foraminifera, copepods, etc., have spines or long hairs to reduce sinking rates.
- ▶ **Benthic**: Benthic organisms live on or are buried in the sea bottom. Tunicates are adapted for a benthic lifestyle by settling to the seafloor in their larval stage and attaching themselves permanently to a hard substrate. As a result, they transform from the non-feeding, planktonic larval stage to an adult that filters food directly out of the water. **Epifaunal** invertebrates are benthic organisms that live on top of the substrate, whether it is sand, mud, or rock. **Infaunal** benthic organisms live between the grains of the substrate. **Sessile** organisms are those that are associated with a substrate and tend to stay attached.
- ▶ **Neuston**: Many invertebrates that have gas floats live at the air/sea interface. This allows them to drift to different areas with both the currents and winds. An example is *Physalia*, the man-o'-war.
- ▶ **Nekton**: Only a few invertebrates are considered truly nektonic organisms, those which can swim well enough to actively move against a current. These nektonic invertebrate organisms are the squid and some large jellyfish; most nekton are vertebrates such as fishes and marine mammals.
- ▶ **Plankton**: Plankton are organisms which either swim weakly or not at all and are passively transported by currents. Most pelagic marine invertebrates are planktonic.

2. MODES OF FEEDING

A wide range of feeding methods exists in the invertebrate realm as a result of the diversification of organisms throughout evolutionary history in both their habitat and lifestyle. Some of the more common feeding styles include:

▶ **Filter-feeding**: Most sessile animals tend to actively filter food directly out of the water column, and have developed complex feeding appendages or gills in order to consume small particles present in the water. (Example: oysters)

▶ **Suspension-feeding**: This term applies to any animal that eats drifting particles or plankton suspended in the water column, whether obtaining it by filter-feeding or other mechanisms. (Example: copepods, tube worms, corals, sponges)

▶ **Deposit-feeding**: Organisms that eat organic matter that has settled to the seafloor by scooping, sieving, or otherwise consuming the substrate and removing the organic matter from the substrate within their digestive system. (Example: spoon worms, lugworms)

▶ **Detrivores**: Organisms that feed on particles of dead or decaying organic matter.

▶ **Herbivores**: Organisms that feed on plant material.

▶ **Carnivores**: Organisms that feed on other organisms.

3. PROTECTION

As with other adaptations, the different ways that marine invertebrates have developed to protect themselves are almost innumerable. Marine organisms not only have to protect themselves against predators, but also against variations in salinity and temperature, water loss, wave action, and other physical factors.

A. Temperature

Marine organisms deal with variations in temperature by developing a variety of protective coverings. Many invertebrates withdraw into shells, while others have developed impermeable outer coverings (e.g., the exoskeleton of crustaceans). Soft-bodied organisms will move into a muddy substrate or deeper water with increasing temperatures. One example of this is the **vertical migration** of planktonic copepods, pteropods, and other organisms downward to deeper waters in the day, and upward back into the surface at night. Such migration is a behavior that these organisms have developed to avoid both temperature stress and predation.

B. Water Loss

Preventing **desiccation,** or the loss of water, is undoubtedly important for marine invertebrates whose bodies are predominantly water. The marine invertebrates most adapted for dealing with desiccation are those along the shorelines, such as rocky or sandy beaches, and in environments where the water level varies with the tides, such as salt marshes and estuaries. In these environments, most organisms either "clam up" or "run and hide" when desiccation becomes a problem. Mussels in the salt marsh will close their shells when the tide is low, opening back up as the tidal current covers the mud. Along the shoreline in the swash zone, the coquina clams will actively burrow into the sand as the water recedes away from the shore with each wave, protecting itself from drying out.

C. Wave Action

Physical action such as waves or other turbulent processes can be harsh to soft-bodied organisms, particularly those that live along the shoreline. Many have therefore developed strategies for coping with high wave activity. Worms tend to burrow into the substrate, and barnacles anchor themselves to rocks using glue that is so strong, many companies have tried to duplicate it (without success). Mussels produce byssal threads, which are very long, strong fibers that anchor them into the muddy substrate.

D. Predation

Marine organisms have developed a tremendous variety of ways for protection against predators, such as spines, poisons, shells, escape methods, etc. Some of the more notable methods are found in the Phylum Echinodermata and Phylum Mollusca. Sea cucumbers will eviscerate their gut when threatened; it is an escape mechanism that allows the sea cucumber to escape (albeit slowly) while the predator feeds on the gut contents left behind. The gut is later regenerated. Sea stars and brittle stars will break off their arms to escape a predator and regenerate the arm later. The radula of cone-shelled gastropods has been modified into a poison dart to not only paralyze their prey, but for protection from their own predators. Some shell-less gastropods, called nudibranchs, feed on corals and jellyfish, saving the cnidarian's nematocysts in its own body to use later for protection against predators.

4. OSMOTIC BALANCE

Marine invertebrates have adapted to the problem of maintaining a proper balance of water and salts in several ways. This regulation of the internal salinity is called the **osmotic balance**. Organisms that control their internal cell concentrations are called **osmoregulators**, actively adjusting the concentration of **solutes** (ions and other materials dissolved into solution) in their body fluids so that they avoid any problems (Figure 8.1). Some osmoregulators can adjust their fluid concentrations to match the external environment; others, such as the blue crab, are capable of varying certain chemicals in their blood, allowing them to live in a wider range of salinities. Some animals do not actively maintain their osmotic balance at all, and thus their internal cell solute concentrations change as the salinity of the water changes. These organisms are called **osmoconformers**, and must stay where the salinity of the water matches that of their fluids. Outside of their narrow salinity range, they experience problems—if placed in freshwater, they swell and burst because of the osmotic flow of water into their tissues. In the open ocean where the salinity changes little, osmoconformers rarely have problems. Organisms that live in highly fluctuating salinity environments such as estuaries and salt marshes must be able to maintain their osmotic balance because the salinity varies widely with the tides and flushing of rivers. In these habitats, many organisms tend to be osmoregulators. Some, such as the blue crab, can switch between osmoregulating at a lower salinity and osmoconforming at higher salinities.

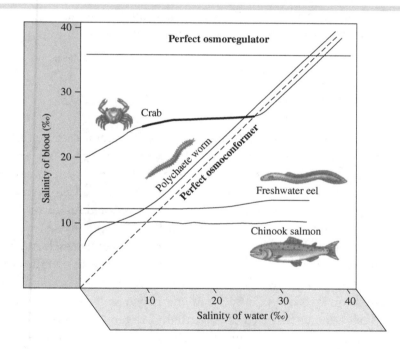

FIGURE 8.1

Osmoregulators vs. osmoconformers. A perfect osmoconformer (dashed line) will match its blood salinity exactly to that of the water, while a perfect osmoregulator will keep its blood salinity the same no matter what the water salinity is. Salmon and freshwater eels are nearly perfect osmoregulators, while a polychaete worm is nearly a perfect osmoconformer. The blue crab is a combination of the two, being only able to regulate its blood salinity within a narrow salinity range (thick black line), while having to osmoconform outside this range. From Marine Biology by P. Castro and M. Huber, 2000. Reprinted by permission of The McGraw-Hill Companies.

5. OTHER ADAPTATIONS

The wide variety of marine organisms and lifestyles provides opportunities for a tremendous array of other adaptations. Some of the more common general adaptations are coloration and camouflage. An interesting example of this is the **chromatophores** of cuttlefish, specialized cells that the animal can actively change to mimic its habitat or indicate its mood. Sensory organs are also important adaptations, such as the light-capturing cells and antennae of copepods. The field of marine invertebrate reproductive strategies is tremendously diverse and organisms exhibit a great range of adaptation. Examples include the multiple larval stages of crustaceans, the reduction and parasitism of the male sex in many molluscs and worms, mass spawning and asexual reproduction by budding in corals, etc.

ADAPTATIONS OF FISHES

As a class, fishes represent the largest group of vertebrates. At present there are about 30,000 known species of fish, and some scientists estimate that there may be over 40,000 species. Fishes are ectothermic animals, typically with backbones, gills, and fins. The study of fishes is called **ichthyology**, and is a fascinating field, as fishes exhibit a tremendous range of adaptation to their habitats.

1. FISH BODY FORMS

Fishes exist in a variety of shapes, forms, and sizes. By carefully examining the body form of a fish (Figure 8.2), you can easily determine how it moves, where it might live, etc. A common body form of fast-swimming, open-water fishes is shown by the tunas and their relatives (Family Scombridae). This streamlined, elongated, torpedo-shaped configuration, with an elliptical cross-section, is called **fusiform**. Many fishes that are not constantly moving but are still capable of quick bursts of speed, such as pinfish, are compressed laterally and are referred to as **compressiform**. This shape allows for greater maneuverability, enabling the fish to make short, quick turns. This body type is often found in fish that inhabit areas where protective cover is available, such as around coral reefs, but is also common in open-water schooling species such as herring. Fish that are flattened dorsoventrally are termed **depressiform**, as with the rays and toadfish. This shape suits the fish for life on the bottom. **Anguilliform** or eel-shaped fishes are adapted for great maneuverability or for burrowing. Other descriptive terms used in describing body form are: **filiform**, for thread-shaped fish such as pipefish for living among reeds; **taeniform**, for the fast and maneuverable ribbon-like shape found in cutlassfishes and gunnels; **sagittiform**, for the arrow-like shapes of pikes and gars needed for fast movement in shallow water; and **globiform**, a body shape used for protection and camouflage as exemplified by the puffers and burrfish.

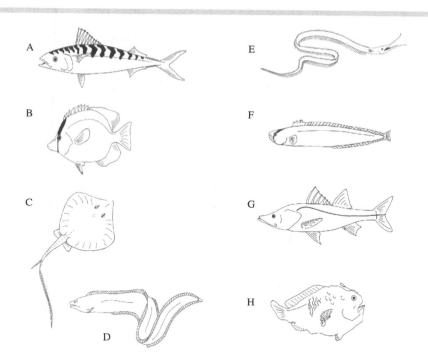

FIGURE 8.2

Representative body shapes in fishes. (A) Fusiform: mackerel (Scombridae); (B) Compressiform: butterflyfish (Chaetodontidae); (C) Depressiform: stingray (Daisyadidae); (D) Anguilliform: moray eel (Muraenidae); (E) Filiform: snipe eel (Nemichthyidae); (F) Taeniform: gunnel (Pholidae); (G) Sagittiform: snook (Centropomidae); (H) Globiform: frogfish (Antennariidae).

Source: Rachel E. G. Kalisperis

There are many variations from these basic shapes, depending on the variability of environments and lifestyles of the fish. Often, the body shape is used as a means of concealment. Some fish resemble leaves or stems of plants in their normal habitats. The sargassum fish *Histrio histrio* is a bizarre-looking fish that looks remarkably like the *Sargassum* algae with which it is associated. The laterally compressed shape of some fish makes them more difficult to detect from a head-on view, while the dorsoventrally compressed shape of some bottom-dwelling fish makes them difficult to distinguish from the seafloor.

2. FISH EXTERNAL FEATURES

A. Gill Structures

Gill openings found in the head region of bony fish include the **operculum**, or gill cover (Figure 8.3). The operculum is not present in more primitive fishes. Agnathans (lampreys and hagfish), sharks, and rays have five or more pairs of external gill apertures through which water exits the gills. **Gill arches** are structures used for support, attachment, and protection of the gills. These are not present in the more primitive agnathans, which have their gills residing in pouches. The higher fishes (cartilaginous and bony fishes, see below) possess gill arches inside the gill structure that provides structural support.

FIGURE 8.3

General exterior anatomy of a fish. (A) Spiny-rayed fish (note the spinous section of the dorsal fin). (B) Soft-rayed fish (note the adipose fin posterior to the dorsal fin).

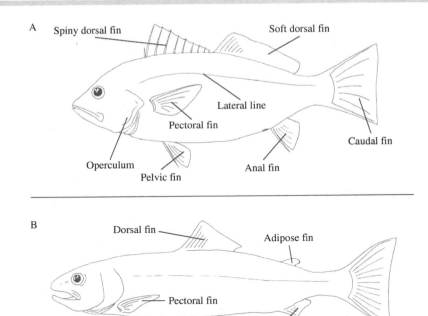

Source: Rachel E. G. Kalisperis

B. Coloration

There is a great variety of coloration in fishes, but general trends can be seen depending on the habitat occupied. Most fish exhibit a pattern of being light-colored on bottom (**ventrally**) and dark on top (**dorsally**), a pattern called **countershading**. This camouflages the fish against the light background of the ocean surface when viewed from below, and against the dark background of the sea bottom or deeper water when viewed from above. Bottom-dwellers and fish living among vegetation are often pale on bottom and intricately marked on top, e.g., flounder, rockfish, etc. The coloration is intended to match their background and provides the fish with a natural camouflage. Open-ocean fishes such as barracuda and herring often have silver sides or iridescent dorsal coloration to blend in with the environment. Mid-water and deep-ocean fishes live in the region of the ocean that is devoid of light, because essentially no sunlight penetrates below about 1000 m. These fishes tend to be either red or black in color, not only to blend in with the surrounding environment but also to absorb any light that might be emitted by other sources. Many deep-sea fishes have **bioluminescent** organs. These organs contain bacteria that produce a chemical reaction when stimulated, giving off a glow of light. Some fishes possess single rows of these organs along their belly, helping to break up their outline when viewed from below. Others have bioluminescent lures that hang from their body to attract prey or potential mates.

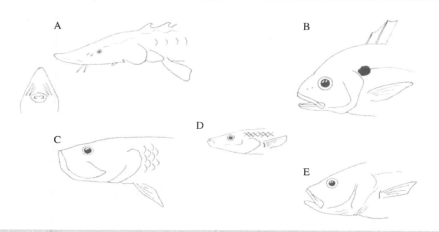

FIGURE 8.4

Mouth positions in fishes. (A) Inferior: sturgeon, with inset (Acipensaridae). (B) Subterminal: spot (Sciaenidae). (C) Superior: tarpon (Elopidae). (D) Superior: killifish (Cyprinodontidae). (E) Terminal: perch (Percidae). See lab text for discussion of differences.

Source: Rachel E. G. Kalisperis

C. Mouth Types

The mouths of fishes are classified according to position (Figure 8.4). The mouth may be **terminal**, situated at the tip of the body, as in trout, freshwater perch, and other estuarine or coastal ocean species. This type of mouth is common in fish that feed in the water column or graze on small invertebrates or plants. A **superior** mouth, which opens upward, is common in surface-feeding fishes such as the mummichog, tarpon, or freshwater killifish. A **subterminal** mouth is common in bottom feeders such as the spot, a common estuarine fish. An **inferior** mouth is situated behind the eye, further back than a subterminal mouth, and is found in bottom feeders such as sturgeon.

D. Fins and Tail Types

Several types of fins are conspicuous features on the fish body (refer back to Figure 8.3). These are supported by skeletal structures and can be either paired or unpaired. The unpaired fins include the **dorsal** (top), **caudal** (tail), and **anal** fin. The paired fins include the **pectorals** and the **pelvic** fins. All of these fins are usually present, but one or more may be absent in some kinds of fish. The fins may be stiffened by structures called rays, which can be either soft and flexible or modified into spines. The cartilaginous fins of sharks and rays are of different structure and origin from those of bony fishes. In bony fishes, true fin spines occur in the dorsal, anal, and pelvic fins. The major function of the fins is for locomotion. Though most fish rely on body movements for general swimming, all depend on fins for movement to some degree. The caudal fin is the most important fin in general propulsion. Fins are also particularly useful for stability and maneuvering. The unpaired dorsal, anal, and caudal fins serve as keels, stabilizing the fish during diving, turning, stopping, etc. In some slower-moving fish with non-streamlined body shapes, such as the pipefish and seahorses, fins are the only means of propulsion.

FIGURE 8.5

Representative shapes of caudal fins. (A) Rounded. (B) Truncate. (C) Emarginate. (D) Lunate. (E) Forked. (F) Heterocercal. See lab text for discussion of differences.

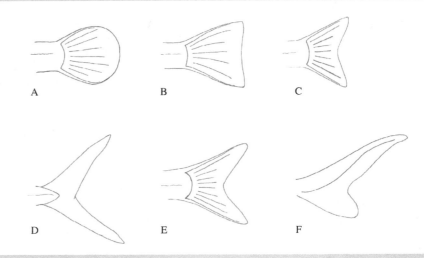

Source: Rachel E. G. Kalisperis

Tail type is an important characteristic used in the classification of the major fish groups. Bony fishes have **homocercal** tails, which is a symmetrical fin shape, as compared to the asymmetrical **heterocercal** structure of shark and sturgeon caudal fins (Figure 8.5F). Bony fishes have several different types of caudal fins (Fig. 8.5 A-E). Those fishes with a crescent or **lunate** shaped caudal tail and a narrow **caudal peduncle** (refer to Fig. 8.3) are generally the fastest, capable of rapid, sustained motion (e.g., tuna, mackerel, swordfish). Many pelagic species have **forked** tails and are constantly on the move, such as sardines. Species with **truncate** or **emarginate** caudal fins may be strong swimmers, but are somewhat slower than those mentioned above. Fishes with **rounded** tail fins tend to be highly maneuverable, but are not capable of producing the propulsion needed to attain the rapid speeds found in fishes with forked or lunate caudal fins.

E. Scales

Scales covering the body are one of the most characteristic features of fish (Figure 8.6). Scales may be lacking, as in catfish, or modified into bony plates or **scutes**, as in sturgeons and sticklebacks. **Placoid** scales are those found on cartilaginous fishes such as sharks; they have a spinous process that gives sharkskin a raspy feel, like a cat's tongue. The diamond-shaped **ganoid** scales of gars are arranged in diagonal rows, embedded in pockets in the skin with a free edge overlapping the next scale. **Cycloid** scales are round and common on many fishes. **Ctenoid** scales, while similar to cycloid scales, differ in that the margin that overlaps the scale next to it has comb-like projections.

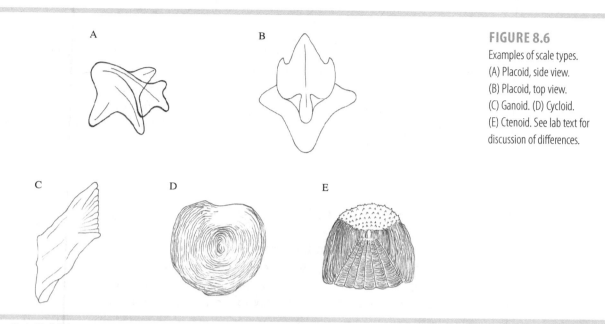

FIGURE 8.6
Examples of scale types.
(A) Placoid, side view.
(B) Placoid, top view.
(C) Ganoid. (D) Cycloid.
(E) Ctenoid. See lab text for
discussion of differences.

Source: Rachel E. G. Kalisperis

3. FISH INTERNAL PHYSIOLOGICAL ADAPTATIONS

A. Swim Bladder

An important adaptation of fishes is the **swim bladder,** present in many bony fishes (but not cartilaginous fishes). This is a gas-filled sac found just above the stomach and intestine, allowing the fish to adjust its buoyancy and keep itself from sinking or rising. Sharks tend to sink because they lack a buoyant swim bladder. To compensate, sharks have large, stiff pectoral fins in combination with the heterocercal tail that provides lift to the body. Because bony fishes have a swim bladder, their pectoral fins are not needed for buoyancy. Thus, these fins have become modified over time for other functions (see Fin Types above).

B. Lateral Line

Fish have a unique sensory organ called the **lateral line.** This feature is found just below the eye, extending from the operculum to the tip of the tail. The lateral line is a system of small canals that run the length of the body, lying in the skin and in the bone (or cartilage) of the head. These canals are lined with sensory cells that are extremely sensitive to vibration, enabling them to detect tiny vibrations in the water resulting from the swimming of other animals or sound waves.

C. Osmotic Balance

Like marine invertebrates, fish must also maintain an osmotic balance with the surrounding seawater. Because the internal fluids of marine fish are much less salty relative to seawater (~14‰ for a fish compared to ~35‰ seawater), the fish will *lose* water to its environment. To make up for this, marine fish drink seawater. The salts present in seawater are excreted by the gills, and the fish release a very small amount of highly concentrated, salty urine (Figure 8.7A).

4. FISH CLASSIFICATION

The evolution and relationships of fishes are still highly debated despite a great deal of research and continuing studies on fish phylogenies. Marine invertebrate phylogenies were presented to you earlier in the semester, and thus we will take time here to present the phylogenetic relationships of marine fishes. You should look for trends in the classification; in particular, the major groups of living fishes are classified by the type of jaw and gills, as well as by the composition of their skeleton:

    I. **Class Agnatha (Cephalaspidomorphi)**—lampreys and hagfishes

        1. Notochord is cord-like

        2. Jawless (**agnathous**)

        3. Main skeleton (vertebrae) cartilaginous

FIGURE 8.7
(A) Marine fishes osmoregulate by drinking seawater, having low urine production, and excreting salt through their gills. (B) Freshwater fishes osmoregulate by not drinking water, having a high output of urine, and absorbing salt through their gills.

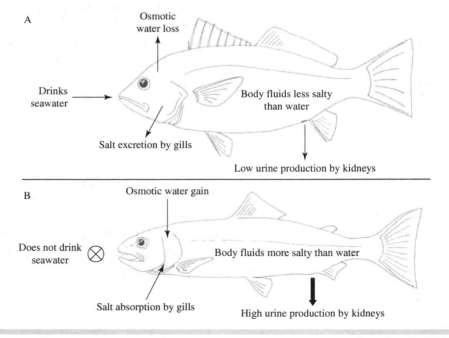

4. No gill arches for support and protection of gills. Instead, the bronchial basket is situated external to the gills, and gills are in pouches.

5. Paired fins absent

6. Single median nostril

7. No scales

II. **Class Chondrichthyes**—cartilaginous fishes

1. Notochord like a string of beads

2. Jawed

3. Cartilaginous vertebrae (some calcification, but no ossification in living forms)

4. Gill arches cartilaginous, internal to gills

5. Paired fins present

   A. Subclass **Elasmobranchii**—sharks, rays, skates

      1. Gills and gill clefts in five to seven pairs

      2. Spiracle present

      3. Placoid scales present or absent

   B. Subclass **Holocephali**—chimaeras

      1. Gills in four pairs, gill clefts a single pair

      2. Spiracle absent

      3. Lacking scales

III. **Class Osteichthyes**—bony fishes

1. Notochord like a cord, string of beads, or like separate beads

2. Jawed

3. Vertebrae bony

4. Gill arches bony, internal to the gills

5. Paired fins present

   A. Subclass **Sarcopterygii**—paired fins have a pronounced lobate, fleshy base with a strong skeletal component. Lungfish.

   B. Subclass **Actinopterygii**—paired fins without pronounced lobate, fleshy base.

EXERCISE 8.1
ADAPTATIONS OF MARINE ORGANISMS

Materials required: Labeled specimens (plastomount or preserved), computer, field or identification guides

Directions

1. Representative specimens exhibiting the adaptations discussed in the lab have been set out in stations and are listed below. Each organism has a particular set of questions associated with it. Answer the questions associated with each specimen.
2. You do not have to start with station number one, but make sure you don't accidentally write the answers of one question in the space for another.

Answer the following questions about each organism:

1. **Lamprey**
   a. The sea lamprey obtains its food by attaching to a fish, rasping a hole through the body wall, and sucking nourishment from the host. What structural features can be seen which indicate an adaptation to this feeding style? _____

   _____

   b. What basic structural differences can be seen which separate the lamprey from higher bony fishes?

   _____

   _____

2. **Black Tip Shark**
   a. The skin of the shark feels rough when rubbed from the posterior to the anterior end, but smooth when rubbed in the opposite direction. What function does this characteristic serve?

   _____

   _____

   b. What function does the heterocercal tail serve for a shark?_____

   _____

3. **Lookdown**
   a. This fish is normally found in open ocean waters. Note the extremely compressed body shape.

   Give a reason for having this type of body shape. _____

   _____

b. What do you think is the purpose of the long filamentous extension of the dorsal fin?

_____

_____

4. **Burrfish**

a. What purpose does the inflated body shape of this burrfish serve? _____

_____

b. What would be a detriment to having this adaptation? _____

_____

5. **Goby**

a. What mouth type does this fish have? _____

b. What body type and caudal fin shape does this fish have? _____

_____

c. Describe an adaptation for protection exhibited by this fish. _____

_____

d. In what type of habitats would this fish live? _____

_____

6. **Lizardfish**

a. What mouth type does this fish have? _____

b. What body type and caudal fin shape does this fish have? _____

_____

c. Describe an adaptation for protection exhibited by this fish. _____

_____

d. In what type of habitats would this fish live? _____

_____

7. **Flounder**

a. Name the body form this fish exhibits. _____

b. Describe what purpose this body form serves for this fish. _____

_____

8. **Moray Eel**

a. Name the body form this fish exhibits. _____

b. Describe what purpose this body form serves for this fish. _____

    _____

9. **Sea Robin**

a. Name the body form this fish exhibits. _____

b. Describe what purpose this body form serves for this fish. _____

    _____

10. **Pompano**

a. Name the body form this fish exhibits. _____

b. Describe what purpose this body form serves for this fish. _____

    _____

11. **Sargassum Fish**

a. Name the body form this fish exhibits. _____

b. Describe what purpose this body form serves for this fish. _____

    _____

12. **Pinfish**

a. Name the body form this fish exhibits. _____

b. Describe what purpose this body form serves for this fish. _____

    _____

13. **Stargazer**

a. Name the body form this fish exhibits. _____

b. Describe what purpose this body form serves for this fish. _____

    _____

14. **Halfbeak**

a. Name the body form this fish exhibits. _____

b. Describe what purpose this body form serves for this fish. _____

    _____

15. **Spot:** Describe the type of caudal fin present and the style of swimming behavior it indicates.

_____

16. **Atlantic Thread Herring:** Describe the type of caudal fin present and the style of swimming behavior it indicates. _____

_____

17. **Spade Fish:** Describe the type of caudal fin present and the style of swimming behavior it indicates.

_____

_____

18. **Mullet**

Examine the specimen and explain what you can discern about its lifestyle from its body form and features. _Explain the reasoning behind your observations._

   a. Type of environment it inhabits: _____

   _____

   b. Swimming ability and maneuverability: _____

   _____

   c. Feeding habits: _____

   _____

19. **Pipefish**

Examine the specimen and explain what you can discern about its lifestyle from its body form and features. _Explain the reasoning behind your observations._

   a. Type of environment it inhabits: _____

   _____

   b. Swimming ability and maneuverability: _____

   _____

   c. Feeding habits: _____

   _____

20. **Toadfish**

Examine the specimen and explain what you can discern about its lifestyle from its body form and features. *Explain the reasoning behind your observations.*

   a. Type of environment it inhabits: _____

   _____

   b. Swimming ability and maneuverability: _____

   _____

   c. Feeding habits: _____

   _____

21. **Octopus**

   a. Name the phylum: _____

   b. Describe a protective adaptation: _____

   _____

   c. Describe at least two other adaptations: _____

   _____

22. **Sea Squirt**

   a. Name the phylum: _____

   b. Describe a protective adaptation: _____

   _____

   c. Describe at least two other adaptations: _____

   _____

23. **Barnacle**

   a. Name the phylum: _____

   b. Describe a protective adaptation: _____

   _____

   c. Describe at least two other adaptations: _____

   _____

24. **Polychaete Worm**

    a. Name the phylum: _____

    b. Describe a protective adaptation: _____

    _____

    c. Describe at least two other adaptations: _____

    _____

25. **Blue Crab**

    a. Name the phylum: _____

    b. Describe a protective adaptation: _____

    _____

    c. Describe at least two other adaptations: _____

    _____

26. **Leech**

    a. Name the phylum: _____

    b. Describe a protective adaptation: _____

    _____

    c. Describe at least two other adaptations: _____

    _____

27. **Pencil Urchin**

    a. Name the phylum: _____

    b. Describe a protective adaptation: _____

    _____

    c. Describe at least two other adaptations: _____

    _____

28. **Oyster**

    a. Name the phylum: _____

    b. Describe a protective adaptation: _____

    _____

    c. Describe at least two other adaptations: _____

_____

29. **Chaetognath**

    a. Name the phylum: _____

    b. Describe a protective adaptation: _____

_____

    c. Describe at least two other adaptations: _____

_____

30. **Common Sponge**

    a. Name the phylum: _____

    b. Describe a protective adaptation: _____

_____

    c. Describe at least two other adaptations: _____

_____

31. **Rainbow Trout:** Describe the method of osmoregulation used by this fish (freshwater):

_____

_____

32. **Atlantic Menhaden:** Describe the method of osmoregulation used by this fish (saltwater):

_____

_____

33. **Atlantic Squid:** Describe the buoyancy adaptation this organism uses:

_____

_____

34. **Perch:** Describe the buoyancy adaptation this organism uses:

_____

_____

35. **Sharpnose Shark:** Describe the buoyancy adaptation this organism uses:

_____

_____

OPTIONAL EXERCISE

The following is a fun exercise that will test your knowledge acquired during the previous lab exercise(s). This exercise may be assigned by your instructor depending on the amount of time available in your lab period.

Team Trivia

Test your lab group's knowledge against the other lab groups in your section. Multiple rounds of questions, composed by your instructor, will be administered one at a time. Questions will be based off material you have learned in lab. The first team to buzz in with the correct answer receives a point. The winning team will be awarded a prize selected by the lab instructor.

Directions

1. Determine your lab group's team name: _____

2. Go to the following website: **https://buzzin.live/**

3. Enter in the Game Code: _____ and your team name.

4. After your instructor administers the question, you will have five seconds to respond. After five seconds you will forfeit your turn and another team will be given the opportunity to answer the question for the point.

Additional Information

If you are interested in further information about the topics in this lab, the following sources will provide you with detailed descriptions, identifying characteristics, and interesting adaptations of marine invertebrates and fishes:

▶ Bone, Q., Moore, R. H. 2008. *Biology of Fishes*. 3rd ed. Taylor & Francis Group: New York.
▶ Long, J. A. 1995. *The rise of fishes: 500 million years of evolution*. Johns Hopkins University Press: Baltimore. 223 pp.
▶ Moore, J. and Overhill, R. 2006. *An Introduction to the Invertebrates*. 2nd ed. Cambridge University Press: Cambridge.
▶ Nelson, J. S., Grande, T. C., and Wilson, M. V. H. 2016. *Fishes of the World*. 5th ed. John Wiley & Sons, Inc.: New York.

LAB 9

DEEP OCEAN ECOSYSTEMS

OBJECTIVES

▶ Understand the adaptations that help marine organisms live in the deep sea.
▶ Explore the hydrothermal vent community and understand why it is such a unique eco-system.
▶ Gain an appreciation for the diversity of organisms in the deep ocean.

POINTS TO PONDER

▶ If there is no phytoplankton performing photosynthesis in the deep sea, then what organisms are at the base of the food chain in this ecosystem?
▶ Why are most deep-sea fishes black or red in color?
▶ Aside from finding a new ecosystem, what was so important about the discovery of hydrothermal vents?

TERMS TO KNOW

abyssopelagic	cold seeps	mesopelagic
Alvin	counter-illumination	pelagic
atmosphere	epipelagic	permanent thermocline
bathypelagic	hadalpelagic	photophores
bioluminescence	hydrostatic pressure	submersible
chemosynthesis	hydrothermal vent	vertical migration

ECOSYSTEMS OF THE DEEP SEA

As you might expect, life for marine organisms in the pelagic realm is very different than in the salt marsh or beach. In contrast to coastal ecosystems, the **pelagic** open ocean lacks the physical structure provided by the seafloor. As a result, there is no place for attachment or protection. The upper region of the pelagic realm, the **epipelagic** zone (Figure 9.1), is defined as the zone from the ocean surface down to approximately 200 meters (the maximum depth where light is adequate for photosynthesis). Of the open ocean regions, this zone has the greatest amount of light and widest range of salinities and temperature. It therefore has the greatest diversity relative to deeper regions.

Amazingly, the deep sea forms the largest habitat on Earth, yet less than 1% of it has been explored. The deeper regions below the epipelagic include the **mesopelagic** (or "twilight zone"), from 200 − 1,000 meters, where there is still faint light but not enough for photosynthesis. The temperature through this zone generally varies less than in the epipelagic, though the mesopelagic is the zone where the **permanent thermocline** occurs (Figure 9.2). Organisms that undergo vertical migration through this region will still encounter large changes in temperature.

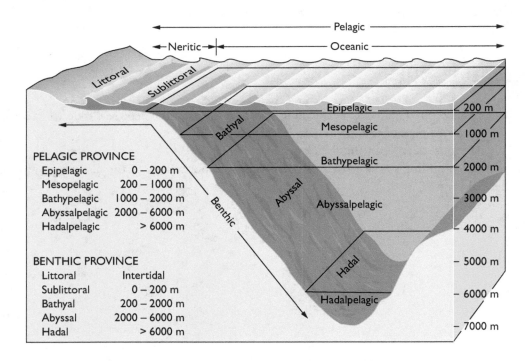

FIGURE 9.1

Zones of ocean habitats. Divisions are primarily a function of water depth. From Paul R. Pinet, *Invitation to Oceanography*, Second Edition, Copyright © 2012: Jones & Bartlett Learning, Burlington, MA. www.jblearning.com. Reprinted by permission.

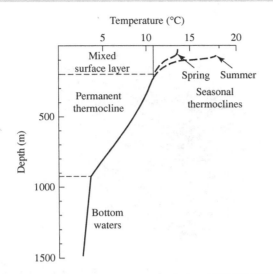

FIGURE 9.2

A general temperature profile showing the seasonal thermocline in the upper surface layer (epipelagic) and the permanent thermocline in the mesopelagic zone. From: Lalli and Parsons, 1997.

Life in the deep sea must adapt to conditions of low or no light, high pressure, reduced food sources, and near-freezing temperatures. Below the epipelagic in the deeper zones, food chains are energy-poor (due to lack of sunlight) and usually begin with bacteria and detritus sinking out of the surface. Mid-water organisms include mesopelagic fish, copepods, shrimp, jellyfish, and cephalopods. Mesopelagic fishes tend to be rather small, with large eyes and mouths. Because of the lack of food in the deep ocean, most mesopelagic animals undergo **vertical migration**, swimming up at night to feed in the surface layers and descending to depth during the day. This migration of organisms is an important pathway for transporting food into deeper water via waste products. Because there is still a faint amount of light in the mesopelagic zone, many organisms in this realm have evolved ways to mask their shape using **photophores**, light-producing cells found primarily on their undersides. These cells produce **bioluminescence** (production and emission of light by an organism) that helps the animal blend in with the dim background light filtering down from the surface, an adaptation called **counter-illumination**.

Below the mesopelagic in the deep sea, sunlight never penetrates. This region is the largest habitat on Earth and contains about 75% of the ocean's water. The deep sea can be divided into several depth zones. The **bathypelagic** zone includes depths between 1,000 and 4,000 meters, and the **abyssopelagic** zone lies between 4,000 and 6,000 meters. Below this is the **hadalpelagic** zone, found only in the deepest ocean trenches. The physical environment of these regions varies little, and organisms in the deep ocean face a constant shortage of food. Less than 5% of the food produced in the upper 100 meters makes it to the deep sea. As a result, animal abundances are very low. Most deep-sea organisms are colored either black or red, because in the absence of light, these colors will blend in with the deep ocean. Organisms in these deep zones do not use bioluminescence for counter-illumination because there is no ambient light to produce an outline. Instead, photophores and bioluminescent

FIGURE 9.3

The humpback anglerfish, also called the black seadevil, is a representative deep-sea fish. Females are typically less than 20 cm (8 in.) in length, and males are only ~3 cm (1 in.) long. The fish possesses a modified dorsal spine with a light organ used as a lure to attract prey. Source: August Brauer, 1906

organs occur along the head and sides, in order to attract prey, mates, or for communication (Figure 9.3). Deep-sea fishes are small, sluggish and sedentary, with flabby muscles, and weak skeletons. Nearly all lack swim bladders. Most adaptations are to maximize locating food and conserving energy. Many have large jaws and are capable of engulfing prey twice their size. Finding a mate is also difficult for these organisms, when such a vast habitat is in total darkness. Many fishes have solved this problem by becoming hermaphroditic, possessing both male and female organs. Some male anglerfishes have become parasitic, living their entire life attached to the female.

Water pressure in the deep sea is one of the many phenomena researchers must contend with when exploring deep ocean sites. What effect does this great depth of water have on organisms living in the deep ocean? Pressure is measured in atmospheres. One **atmosphere** is equal to the weight of Earth's atmosphere at sea level, about 14.6 pounds per square inch. Thus at sea level, each square inch of your body is subjected to a force of 14.6 pounds. The pressure increases about one atmosphere for every 10 meters of water depth. At a depth of 5,000 meters the pressure will be approximately 500 atmospheres or 500 times greater than the pressure at sea level. As expected, high **hydrostatic pressure** (from the weight of the water column overhead) affects air-filled organs, such as fish swim bladders. When deep-sea fishermen pull up fish from the mesopelagic and bathypelagic zones, the swim bladder has often expanded out of its mouth and ruptured, killing the fish.

DEEP-SEA CORALS

Advances in ocean exploration have uncovered coral gardens in the dark ocean depths. Deep-sea corals (or cold-water corals), live up to 6,000 meters below sea level. They can survive in waters as cold as −1°C and are able to live in many different places around the world. Instead of rock-like reefs, these cold-water corals form groves of tree, feather, column or fan shapes. They do not require sunlight as a source of nutrition, but instead feed on microscopic organisms that flow in the ocean currents. As of 2019, more than 3,300 species of deep-sea corals have been identified and the numbers keep climbing. Currently, some of these coral colonies are the oldest marine organisms on record, with one colony in Hawaii measured to be about 4,265 years old.

CHEMOSYNTHETIC ECOSYSTEMS

Hydrothermal Vents

Before 1977, deep-sea vent communities were unknown. These ecosystems were discovered using the tiny **submersible** craft *Alvin*, operated by Woods Hole Oceanographic Institute. While diving nearly 2,400 meters on the East Pacific Rise near the Galapagos Islands, the submersible and its scientists happened upon a **hydrothermal vent**, the first ever seen (Figure 9.4). Completely isolated from sunlight, entire com-

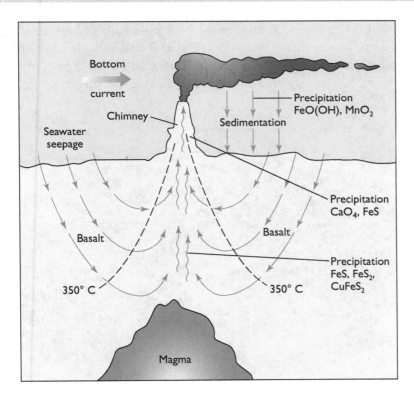

FIGURE 9.4

Heat from a shallow magma chamber causes seawater to flow through the fractured seafloor basalt. Out of this rock are leached metals that are discharged with the hot fluid from a chimney. When the fluids cool, chemical reactions precipitate sulfide and other minerals. From Paul R. Pinet, *Invitation to Oceanography*, Second Edition, Copyright © 2012: Jones & Bartlett Learning, Burlington, MA. www.jblearning.com. Reprinted by permission.

munities of organisms were found to live around these vents. Hydrothermal vents form at mid-ocean ridges (Figure 9.5), where two oceanic plates pull apart and rising magma cools to form seafloor. In these areas, extremely hot (upwards of 400°C) mineral-rich fluid flows out of cracks in the surface of the seafloor. The hot fluid flows into very cold water, usually ~2°C, and cools down quickly. The cooled minerals in the fluid settle around the vent opening, creating chimney-like formations.

These hydrothermal vents have abundant life for as long as the vents remain active, which is only about one to two years. In fact, more than 300 species live around the vents and are unique to this type of environment. Creatures such as tubeworms, fish, crabs, clams, anemones and bacteria have learned to survive the complete darkness, the extremely hot vent water and the tremendous water pressure. At such depths, photosynthesis cannot occur, thus the base of the food chain cannot be phytoplankton or other algae. So what other option is there? A unique ecosystem has evolved at these depths in the absence of sunlight, where animals rely on bacteria that can convert sulfur found in the vent's fluids into energy through **chemosynthesis**. The chemical reaction that occurs uses the available dissolved oxygen and carbon dioxide in the water and hydrogen sulfide emitted from the vents to produce organic matter:

$$H_2O \ + \ O_2 \ + \ CO_2 \ + \ H_2S \longrightarrow CH_2O \ + \ SO_4^{2-} \ + \ 2H^+$$

The animals that live around hydrothermal vents essentially make their living from chemicals coming out of the seafloor in the vent fluids! This energy is then transferred through the food chain as larger organisms feed on the chemosynthetic bacteria or feed on the animals that eat the bacteria. Because they are so localized in

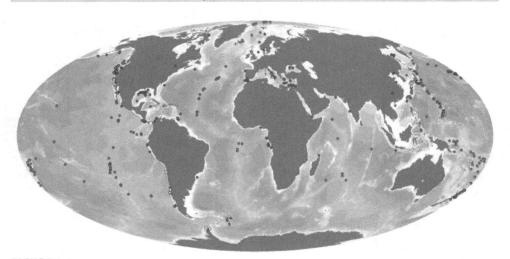

FIGURE 9.5

Global distribution of hydrothermal vent (Red dots), cold seep (blue dots), and whale fall (yellow dot) sites that have been studied with respect to their fauna. Source: Deep-Water Chemosynthetic Ecosystem Research during the Census of Marine Life Decade and Beyond: A Proposed Deep-Ocean Road Map by Christopher R. German, Eva Ramirez-Llodra, Maria C. Baker, Paul A. Tyler, and the ChEss Scientific Steering Committee. Published August 4, 2011. https://doi.org/10.1371/journal.pone.0023259. https://creative-commons.org/licenses/by/4.0/

nature, hydrothermal vents have high biomass relative to the very sparse distribution of animals outside of vent areas, where animals are primarily dependent on food falling through the water column.

Perhaps the most striking creatures of the hydrothermal vent community are the tube worms. These worms are so unusual that they were originally placed in their own phylum (Vestimentifera), but recent molecular and genetic analyses have shown them to be related to annelids and thus are now a family within Phylum Annelida. Tube worms are encased in leathery tubes with a plume of bright red tentacle-like respiratory filaments protruding from the open end. They lack a mouth or even a digestive tract; instead, they absorb nutrients from chemosynthetic bacteria that live inside their bodies. The largest tubeworms measured ~1.5 meters long with tubes up to 3 meters in length. Tubeworms may also be the fastest growing animals in the world, growing at rates up to 85 cm/yr!

Cold Seeps

Cold seeps are another deep-sea environment in which hydrocarbon-rich chemicals contribute energy to the ecosystem. First discovered in 1983 off Florida's Gulf coast, these ecosystems have been found throughout the world's oceans on continental margins and in areas of tectonic activity, where cracks and fissures in the seafloor allow hydrocarbon-rich fluids such as methane and hydrogen sulfide to bubble up from petroleum deposits below the seafloor (Figure 9.5). Several types of cold seeps exist, including brine pools, gas hydrate seeps, and methane seeps. Methane seeps are considered one of the most geologically diverse and widely distributed systems explored to date, and new sites are being discovered every year.

Cold seeps have several important differences in contrast to hydrothermal vents. As their name implies, cold seeps develop at temperatures close to the surrounding waters. Because the seeps do not depend on heat from magma plumes (as with vents), they function differently than hydrothermal vents and are more stable and long-lasting. Similar to hydrothermal vent ecosystems, cold seep environments have low diversity despite the high biomass present. Bacteria metabolize the methane or hydrogen sulfide for energy, contributing this energy source to the ecosystem through a symbiotic relationship with the other organisms. Mats of chemosynthetic bacteria and mussel beds first colonize and dominate cold seep communities, because they are more tolerant to the chemicals released from the seafloor. Once the hydrocarbon emissions slow, other organisms such as giant tubeworms move into the region. The clams along with giant tubeworms (of a different species than at hydrothermal vents) are all nourished by symbiotic bacteria in their tissues. Because of the low temperature and stability of the cold seeps, worms in these environments can live for more than 200 years.

Brine pools are a type of cold seep with highly saline fluids escaping the seafloor. Brine is a solution many times saltier than seawater (>50 ppt), and are formed by a process called *salt tectonics,* in which large underground salt deposits dissolve and move as a result of tectonic plate motion or faulting. Because the brine is significantly denser than the surrounding water, the brines flow out of the seafloor and form pools and lakes underwater with a distinct surface and shoreline; one in the Gulf of Mexico is up to 20 km long! This brine water can become saturated with methane bubbling up from subsurface deposits, and in such cases, mussels with symbiotic bacteria living in their gills are usually found surrounding the shoreline.

ADDITIONAL INFORMATION

If you are interested in further information about the topics in this lab, the following websites provide a tremendous amount of information, images, video clips, and links concerning the deep-sea ecosystem and hydrothermal vents:

- ▶ Project NeMO: New Millennium Observatory. NOAA Pacific Marine Environmental Laboratory Vents Program. http://www.pmel.noaa.gov/vents/nemo1998/project.html.
- ▶ Yancey, P.H. "DEEP-SEA Pages." http://people.whitman.edu/~yancey/deepsea.html
- ▶ Yancey, P.H. "DEEP-SEA Pages: Mesopelagic (Midwater) Animals." http://people.whitman.edu/~yancey/midwater.html

EXERCISE 9.1
EXPLORATION OF DEEP OCEAN ECOSYSTEMS

Materials needed: Computers with internet access

The NOAA Office of Ocean Exploration and Research (OER) is a federal organization dedicated to exploring Earth's oceans. OER's research expeditions explore previously unknown areas to observe and document all aspects of the ocean, including unknown oceanographic features, ecosystems, marine resources, archaeological sites, and much more. Since 2009, expeditions around the globe have been conducted using the NOAA research vessel (R/V) *Okeanos Explorer,* which deploys a ROV (Remotely Operated Vehicle) to explore the undiscovered depths of the ocean.

Directions

Part 1 of this lab is an exploratory-based exercise that takes you through an exploration of the different deep-sea ecosystems. OER's website is designed to provide the public with information on the program's ongoing activities and discoveries. In this activity you will explore the OER website, watching videos and answering questions about what their explorations have found. Go through the questions in order, as they take you step-by-step through the websites and interpretation.

Chemosynthetic Oases

1. Visit the following site:

    http://oceanexplorer.noaa.gov/welcome.html

    Click on the Menu link, and then select "Expeditions" – "By Year" – "2017". Select "Gulf of Mexico 2017", then scroll down to the "Image and Video Gallery" and select it. Find the "Chemosynthetic Oases" video, watch the video, and answer the questions below.

2. a. What type of organism act as the base of the food web at these communities?

    _____

    b. What chemical does it use to make energy? _____

    _____

3. What are foundation species? _____

    _____

4. What attracts other species to live in these areas? _____

    _____

5. Why is it important to explore chemosynthetic habitats (what are their ecological services)?

_____

_____

Cold Seeps

Next go to "Expeditions" – "By Year" – "2018". Select "Gulf of Mexico 2018", then scroll down to the "Image and Video Gallery" and select it. Find the "Exploring Seepscapes" video, watch the video, and answer the questions below.

6. What are cold seeps? _____

_____

7. What two types of fluids are found at cold seeps? _____

8. How do brine pools form? _____

_____

9. What is a gas seep? _____

10. How long can these seeps last? _____

Deep-Sea Corals

Return to "Expeditions" – "2017" – "Gulf of Mexico 2017", scroll down to the "Image and Video Gallery" and select it. Find the "Architects of the Deep" video, watch the video, and answer the questions below.

11. How are deep-sea corals different from shallow-water corals? _____

_____

12. How do deep-sea corals survive in the deep ocean? _____

_____

13. List three reasons why deep-sea corals are so important: _____

_____

Hydrothermal Vents

Click on "Expeditions" – "By Year" – "2016". Select "2016 Deepwater Exploration of the Marianas", then scroll down to the "Image and Video Gallery" and select it. Scroll down to the "Oases of Life" video, watch the video and answer the questions below.

14. What is the purpose of monitoring the deep-sea ecosystem over time? _____

_____

15. a. As the ROV moves around this vent, what particular characteristics do you notice about the water in the deep ocean? _____

_____

b. Describe the activity of these vents in your own words: _____

_____

_____

c. Many of the same species of deep-sea vent organisms are found distributed along all mid-ocean ridges (that scientists have found so far). What type of dispersion pattern is this called? _____ How might these organisms spread from site to site?

_____

Scroll down and locate "Dive 11: Hydrothermal Vent" video and watch it.

16. a. At what depth is this vent? _____

b. What type of "smoke" does this chimney produce? _____

c. How tall is this vent? _____

Figure 9.6 below represents a hydrothermal vent system. Trace the movement of seawater through the vent system in the diagram and answer the following questions. You may also browse the following website for additional help:

http://www.divediscover.whoi.edu/vents/chemistry.html

17.   What enables the seawater to pass through the rock? _____

18.   What happens to the temperature of seawater as it passes down through the rock?

_____

19.   As the water passes down through the rock, it picks up elements. What are they? _____

_____

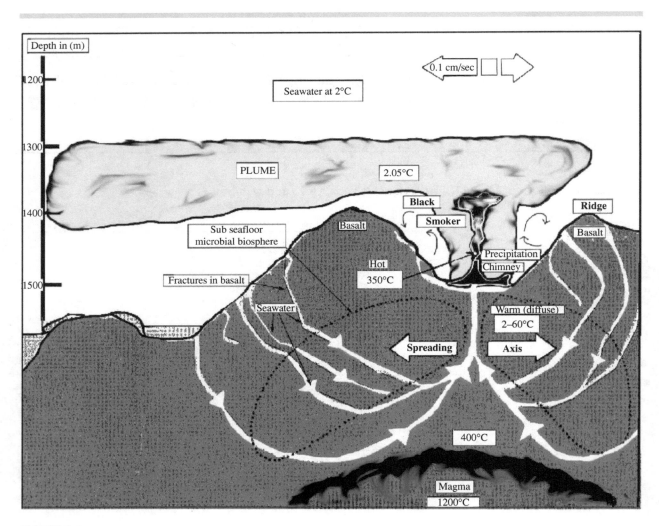

FIGURE 9.6

Diagram of a hydrothermal system. Source: http://www.pmel.noaa.gov/vents/nemo1998/curriculum.html.

Next, visit the following site and watch the video to answer the questions below:

https://ocean.si.edu/ocean-videos/hydrothermal-vent-creatures

20. What is the temperature of the vent water? _____

21. Toxic chemicals spew from hydrothermal vents. List three toxic chemicals and the associated concentrations at which they can be found at vents: _____

    _____

22. The abyssal depths had always been regarded as a desert, incapable of supporting life with no sunlight available for photosynthesis. Yet to everyone's surprise, scientists found abundant life on the seafloor, especially around the hydrothermal vents. Describe briefly what you view as some of the extreme environmental conditions in which these organisms live.

    _____

    _____

    _____

23. Based on the information provided in the websites, what role do the chemosynthetic bacteria play in these deep ocean environments? _____

    _____

    _____

EXERCISE 9.2
MARINE ADAPTATIONS TO THE DEEP SEA

1. While relatively little is known about mid-water and deep-sea organisms, scientists have begun exploring these realms and discovering tremendous amounts of new information and new species. The following websites have some spectacular images, videos, interactive activities, and information to help you answer the next set of questions:

   a. http://www.seasky.org/deep-sea/deep-sea-menu.html

   b. http://www.montereybayaquarium.org/animals-and-experiences/exhibits/mission-to-the-deep

   c. https://ocean.si.edu/ecosystems/deep-sea

   d. http://ocean.si.edu/deep-sea-corals

2. Bioluminescence is an adaptation exhibited by a wide range of marine organisms. Answer the following questions on bioluminescence:

   a. Some organisms squirt ink that illuminates as it comes in contact with the dissolved oxygen of seawater. Suggest the purpose for this biochemical adaptation.

   _____

   _____

   b. On the lantern fish, photophores are found mainly along their belly. What is the reason for having the photophores concentrated on their ventral surface?

   _____

   _____

   c. The photophores on a given fish species exhibit a species-specific pattern (a pattern unique to that species). Of what use might this specific photophores pattern be to the fish? _____

   _____

3.   From the websites, select five organisms from the deep ocean. List one unique adaptation they have
     developed allowing them to survive in the deep sea and that differs from the other deep-sea organ-
     isms you chose. Provide the function of each adaptation.

Organism	Adaptation for the Deep Sea	Function

4.   Many fish living in the deep ocean do not have scales. What is the purpose for having this particular
     adaptation? _____

     _____

5.   Why do many deep-sea fish have poorly developed muscles? _____

     _____

6. In many deep-sea species, the male becomes totally dependent on the female, and only its reproductive system continues to fully function. Why would this be a helpful adaptation for the deep-sea environment? _____

_____

_____

7. Why do vertical migrators move to shallow water at night and down to deeper regions in the day? List three reasons that you can think of. _____

_____

_____

_____

_____

_____

_____

APPENDIX A

HOW TO USE A MICROSCOPE

Since the majority of the ocean realm is populated by tiny plankton and single-celled bacteria, it is important to know the correct procedure for using a microscope. Furthermore, microscopes are very expensive to replace, and knowing how to operate one correctly will help to prevent accidental damage to the equipment. The operation of the microscope is provided for you in sequential steps. Please refer to the figure below as you read the following directions.

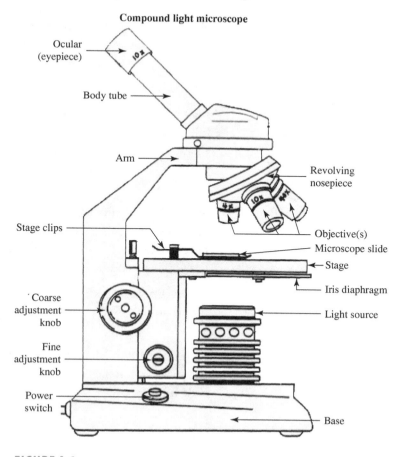

Compound light microscope

Ocular (eyepiece)
Body tube
Arm
Stage clips
Coarse adjustment knob
Fine adjustment knob
Power switch
Revolving nosepiece
Objective(s)
Microscope slide
Stage
Iris diaphragm
Light source
Base

FIGURE A.1

Parts of the compound light microscope. Courtesy of www.southwestschools.org

DEFINITIONS TO KNOW AND PARTS TO RECOGNIZE

Arm:	Supports the body tube
Base:	Supports the microscope; houses the light source or mirror (on older models)
Body tube:	Contains the ocular and objective lenses
Coarse adjustment knob:	A small turn of this focusing knob makes a relatively large change in the distance between the objective and the slide
Fine adjustment knob:	A small turn of this focusing knob makes only a small change in the distance between the objective and the slide
Iris diaphragm:	An adjustable aperture below the stage that functions to regulate the amount of light passing through the specimen
Light source:	A light bulb attached to the base underneath the stage
Magnification:	How much the object is enlarged; it is determined by multiplying the power engraved on the side of the objective lens used by the power of the ocular lens (usually 10X); (for a 45X objective, the magnification would be 450X the natural size)
Nosepiece:	Revolving turret with attached objectives
Objectives:	Each objective is a system of magnifying lenses; some microscopes have three objectives, and others have four (magnifications are: **scanning power** = 3–4X; **low power** = 10X; **high power** = 43-45X; **oil immersion** = 90–100X)
Ocular (eyepiece):	System of magnifying lenses; the usual magnification of the ocular itself is 10X, though some oculars do not have a magnification lens inside; if present, the magnification is usually written on the outside of the ocular
Resolution:	The ability to see adjacent forms as distinct from each other; limited by the wavelength of light
Stage:	Flat platform that supports the slide; some microscopes have a mechanical clip mechanism for moving the slide, others have simple stage clips that hold the slide in place over the light source
Stage clips:	Metal clips that hold the microscope slide in place on the stage

DIRECTIONS FOR OPERATING A MICROSCOPE

1. Always carry the microscope with one hand on the **arm** and one hand under the **base**. Carry it close to your body and handle it gently as you carry it to your station. Place the microscope on your table with the arm toward you. Remove the cover and plug the microscope in.

2. Familiarize yourself with the parts of your microscope, as not all scopes are made alike.

3. Before using your microscope you should always clean the lenses as follows: Use the **coarse adjustment knob** to raise the **nosepiece** to its maximum height above the stage. Wipe the **ocular** and all **objectives** with lens paper only.

4. *Always* start and end with the lowest power. Rotate the **nosepiece** so that the shortest (lowest magnification) objective is lined up with the body tube. It will click into place. This lens magnifies an object 4 times its original size (4X).

5. Turn on the microscope light and place the specimen to be viewed over the hole in the **stage**, directly over the light source. You may have to open the stage clip lever on the stage or lift up the **stage clips** themselves, depending on the type of microscope. Gently close the stage clips—do not let the stage clip lever snap back onto the slide, as it will break the glass. On some scopes you can move the stage clip mechanism and slide together beneath the objective field by using a series of knobs underneath the stage.

6. *Before* looking through the microscope, lower the 4X objective down to about 5 mm above the specimen slide (nearly its lowest point) using the **coarse adjustment knob**. Watch the progress of the objective from the side so as not to drive the lens through the cover slip and glass slide.

7. Look through the **ocular** lens (eyepiece) and slowly raise or lower the objective lens using the coarse focus knob until you see an image. Center the object in your field of view and bring the image into better focus by adjusting the **fine adjustment knob**. Only a slight adjustment should be necessary. The specimen will be in focus when the low power objective is close to the lowest point, so start there and focus by slowly raising the lens. If you can't get it at all into focus using the coarse knob, then switch to the fine focus knob.

8. Once you have the specimen centered and in focus on your lowest powered objective, then, *without* changing the focus knobs, rotate the nosepiece to the next greater powered objective (either 10X or 40X depending on the scope). Watch from the side as you rotate the objective to make sure it doesn't come in contact with the slide. If it does, you should first lower the stage slightly using the coarse adjustment knob.

9. If you don't center the specimen in the low power field of view, you will lose it when you switch to a higher power. If the specimen is still not in your field of view after rotating to the higher objective, return to the lower power objective and re-center the object as described in steps 6 & 7.

10. Once you have it on high power, remember that you *only* use the fine adjustment knob. Using the coarse adjustment will run the objective into the slide, breaking the glass and possibly damaging the objective lens.

11. When you have finished examining the specimen, *always* rotate the lowest power objective over the specimen *first*. Then lower the stage by turning the coarse adjustment knob as far down as it will go, allowing you to safely remove the slide.

IMPORTANT NOTES

► You can further adjust the focus of each individual ocular (in bifocal microscopes) by twisting the ocular right or left to bring the individual lens into focus.

► If it is available on your microscope, you can adjust the amount of light passing through the specimen by opening or closing the **iris diaphragm** (as you move the lever, look down through the hole in the stage to see the light get brighter). As you adjust the diaphragm, more detail is visible when you allow in less light—too much light will give the specimen a washed-out appearance.

► You should *never* use the highest power objective (usually 100X, and has the longest tube) as this is an oil immersion lens and requires special training and supplies to use. We will never look at organisms small enough to require viewing under the oil immersion lens.

► If you wear glasses, take them off; if you see only your eyelashes, move closer. It may help to close or cover your other eye if you are using a single-ocular microscope.

► If you see a dark line that goes part way across the field of view, try rotating the eyepiece. That dark line is a pointer that can be used to point out something to your lab partner or your instructor.

APPENDIX B

A BRIEF GUIDE TO STATISTICAL ANALYSIS USING MICROSOFT EXCEL®

(Microsoft Excel® 2016 version for Mac and PC)

Microsoft Excel® is a powerful program that allows you to easily manipulate large data sets. In its simplest form, you can use Excel to organize, sort, and analyze your data. As you become more advanced, you can use Excel to graph and even measure statistical significance. As future marine scientists, it is essential that you become familiar with programs such as these so that you can examine your data and determine the validity of your results. While you read through this guide, follow along in an Excel spreadsheet so that you can visually see exactly what the guide is referring to.

Open a new spreadsheet in Excel on your computer, click on Microsoft Excel (green icon with an X in the middle) and choose the 'Blank workbook' option. Check to make sure you have the 'Data analysis toolpak' installed by clicking on the 'File' tab and click 'options'. Then choose the 'Add-Ins' option and click 'Go'. A new pop-up window will appear and make sure the 'Analysis ToolPak' option is checked and click 'ok'. This will allow you to run the desired statistics in Excel. [Mac users: click on 'Data' in the toolbar, then click 'Excel AddIns'. Check the box for 'Analysis ToolPak' and click 'ok'.]

INPUT THE DATA

To get started, enter in the data from the table below. This data was recorded at noon every day for a week from two sites in the North Inlet Estuary on the coast of South Carolina.

You can find the symbol for degrees by clicking the 'Insert' tab and then choose 'Symbol'. If it's not in the immediate choices choose 'More Symbols...' to look for it. (Note: if you double-click on the line between the rows/columns it will automatically resize the spacing to fit the text.) We are going to calculate the mean (Avg.), standard deviation (SD), and standard error (SE), so go ahead and label cells for these calculations (see cells A19–A21 in Figure B.1).

Observation #	Date Collected	Temperature (°C)	Dissolved Oxygen (mg/L)	Location
1	5/01/14	22.2	6.0	Oyster Landing
2	5/02/14	20.2	6.0	Oyster Landing
3	5/03/14	18.9	6.6	Oyster Landing
4	5/04/14	20.0	6.0	Oyster Landing
5	5/05/14	21.3	5.5	Oyster Landing
6	5/06/14	22.8	5.1	Oyster Landing
7	5/07/14	24.1	4.1	Oyster Landing
8	5/01/14	22.2	6.1	Clam Bank
9	5/02/14	19.1	6.6	Clam Bank
10	5/03/14	18.2	7.4	Clam Bank
11	5/04/14	18.6	7.0	Clam Bank
12	5/05/14	19.8	7.1	Clam Bank
13	5/06/14	22.2	5.8	Clam Bank
14	5/07/14	22.8	4.9	Clam Bank

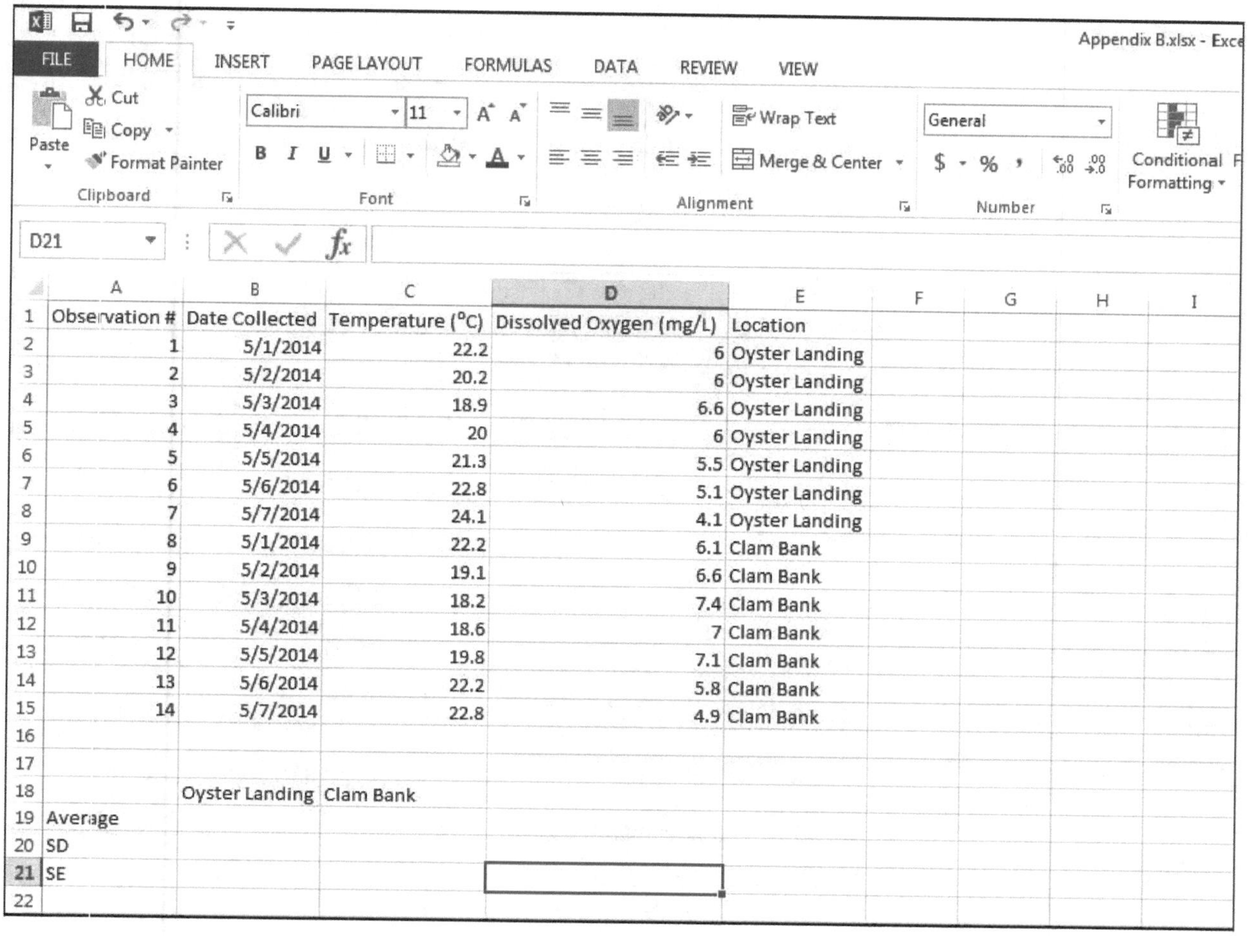

	A	B	C	D	E
1	Observation #	Date Collected	Temperature (°C)	Dissolved Oxygen (mg/L)	Location
2	1	5/1/2014	22.2	6	Oyster Landing
3	2	5/2/2014	20.2	6	Oyster Landing
4	3	5/3/2014	18.9	6.6	Oyster Landing
5	4	5/4/2014	20	6	Oyster Landing
6	5	5/5/2014	21.3	5.5	Oyster Landing
7	6	5/6/2014	22.8	5.1	Oyster Landing
8	7	5/7/2014	24.1	4.1	Oyster Landing
9	8	5/1/2014	22.2	6.1	Clam Bank
10	9	5/2/2014	19.1	6.6	Clam Bank
11	10	5/3/2014	18.2	7.4	Clam Bank
12	11	5/4/2014	18.6	7	Clam Bank
13	12	5/5/2014	19.8	7.1	Clam Bank
14	13	5/6/2014	22.2	5.8	Clam Bank
15	14	5/7/2014	22.8	4.9	Clam Bank
16					
17					
18		Oyster Landing	Clam Bank		
19	Average				
20	SD				
21	SE				
22					

FIGURE B.1

Excel spreadsheet setup for example data. Source: Emily G. Baumann. Used with permission from Microsoft Excel.

DATA CALCULATIONS

You can manipulate your data by inputting simple equations into Excel in a specific way. The '=' sign tells the program that an equation is about to be entered. For example, you want to determine the mean (average) of dissolved oxygen from Oyster Landing, SC, for the first week of May. Go to the empty cell, B19, and type '=average(D2:D8)' (do not put any spaces into the formula and remove the apostrophes), and hit 'Enter'. This selects the cells with the Oyster Landing data and tells the program to take the average of them. The number, 5.614286, should appear in that cell. However, if you look above the spreadsheet in the Formula Bar (*fx*), the equation is displayed for you instead of the value. Now, in cell C19, calculate the mean of the dissolved oxygen at Clam Bank for the same week. You should get 6.414286.

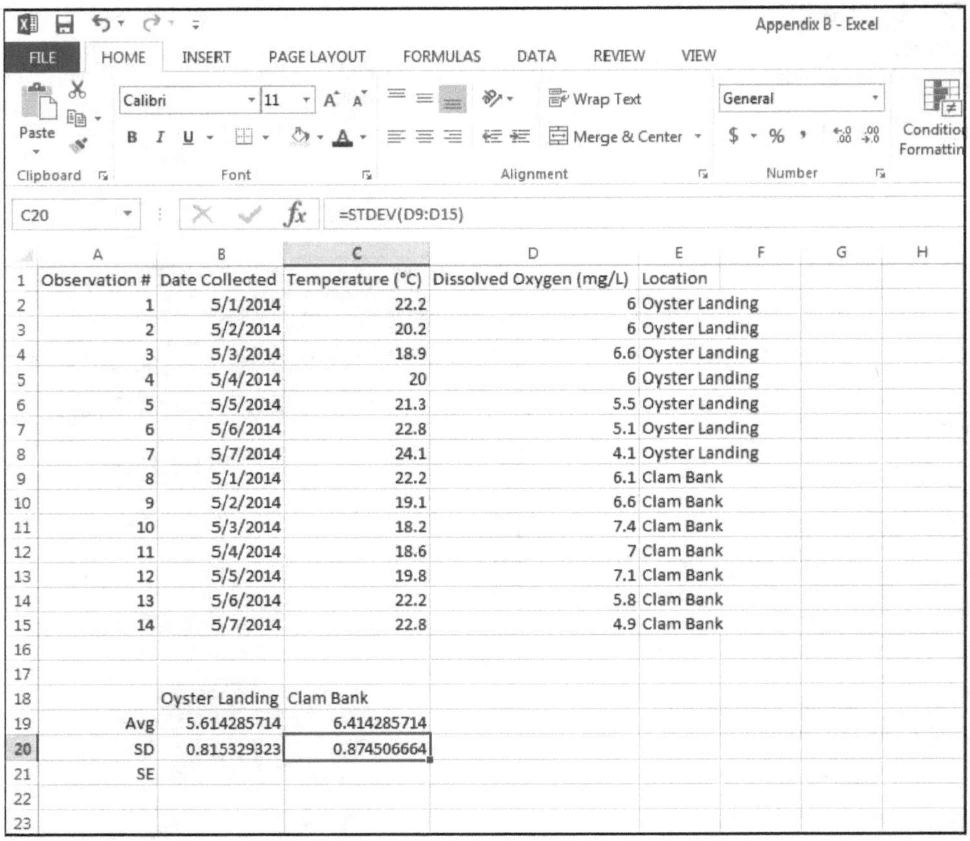

	A	B	C	D	E	F	G	H
1	Observation #	Date Collected	Temperature (°C)	Dissolved Oxygen (mg/L)	Location			
2	1	5/1/2014	22.2	6	Oyster Landing			
3	2	5/2/2014	20.2	6	Oyster Landing			
4	3	5/3/2014	18.9	6.6	Oyster Landing			
5	4	5/4/2014	20	6	Oyster Landing			
6	5	5/5/2014	21.3	5.5	Oyster Landing			
7	6	5/6/2014	22.8	5.1	Oyster Landing			
8	7	5/7/2014	24.1	4.1	Oyster Landing			
9	8	5/1/2014	22.2	6.1	Clam Bank			
10	9	5/2/2014	19.1	6.6	Clam Bank			
11	10	5/3/2014	18.2	7.4	Clam Bank			
12	11	5/4/2014	18.6	7	Clam Bank			
13	12	5/5/2014	19.8	7.1	Clam Bank			
14	13	5/6/2014	22.2	5.8	Clam Bank			
15	14	5/7/2014	22.8	4.9	Clam Bank			
16								
17								
18		Oyster Landing	Clam Bank					
19	Avg	5.614285714	6.414285714					
20	SD	0.815329323	0.874506664					
21	SE							
22								
23								

FIGURE B.2
Values for averages and standard deviations. Note the equation is displayed in the Formula Bar (*fx*), but the value is displayed in the cell. Source: Emily G. Baumann. Used with permission from Microsoft Excel.

You can also use Excel to find the **standard deviation** (SD). The standard deviation is calculated by taking the square root of the variance. The **variance** is the mean of squared deviations of observations from their arithmetic mean but is rarely used as a descriptive statistic. However, many statistical tests use variance in their calculations. Use Excel to calculate the SD of your measurements, by typing '=stdev(D2:D8)'. You should obtain a value of 0.81. Now calculate the SD for Clam Bank (see Figure B.2).

Now, let's use Excel to calculate the **standard error** of the dissolved oxygen at Oyster Landing. Go to cell B21 and type the following '=B20/(sqrt(7))' and press 'Enter'. This tells Excel to take the SD that you already calculated in cell B20 and divide (/) it by the square root of your sample size of 7. Now calculate the SE for Clam Bank in cell C21 (see Figure B.3). Your value should be 0.33.

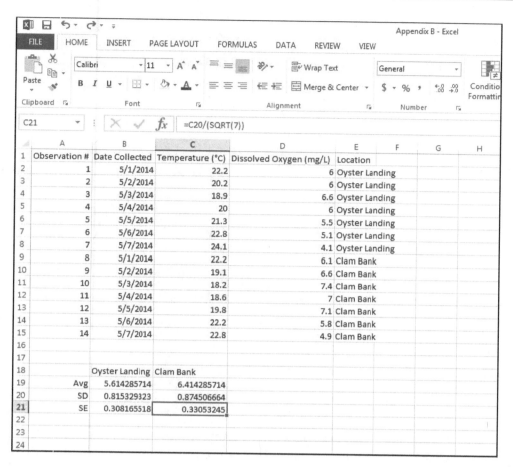

FIGURE B.3

Values for standard errors. Again, note the equation is displayed in the Formula Bar (*fx*), but the value is displayed in the cell. Source: Emily G. Baumann. Used with permission from Microsoft Excel.

PLOTTING THE DATA FOR COMPARISON

When comparing data, figures are an excellent tool to display your results. If you're comparing variables, a bar graph is a useful aid. Let's compare the means of dissolved oxygen for each sample site. First, highlight the cells containing the sample sites and averages and click the 'Insert' tab on the toolbar. Choose 'Column' from the list of charts and click on the first option under the '2-D Column' choices labeled 'Clustered Column'. A bar graph will appear. Click on the 'Switch Row/Column' icon on the toolbar. This creates two separate data series (different colored bars). Delete the '1' from the X-axis by clicking on it and pressing the 'Delete' key. Click on the 'Chart Elements' icon (green plus sign + at the top right corner of the graph) and check the box for 'Legend'. Then check the box for 'Axes Titles' and label the X and Y-axes, and don't forget the units! (See Figure B.4.) [Mac users; you can either right-click the data points on the graph or you can click on 'Chart Design'—'Add Chart Element'—'Axes Titles'.]

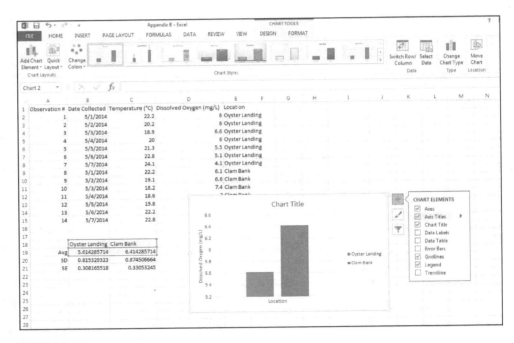

FIGURE B.4

Adding and editing X-axis and Y-axis titles under the Chart Elements icon tab. Source: Emily G. Baumann. Used with permission from Microsoft Excel.

Next we will add the error bars to show the standard error of both the means. Click on the 'Chart Elements' icon, choose the 'Error Bars' and click on the 'More Error Bars Options...'. Choose the 'Oyster Landing' option from the pop-up window and click 'OK'. Choose the 'Custom:' choice and click on 'Specify Value'. Click the icon for 'Positive Error Value' and highlight the SE cell under Oyster Landing (refer to Figure B.5) and hit the 'Enter' key. Now do the same procedure for 'Negative Error Value' and hit 'Enter'. Click 'OK' and then click 'Close'. Now go back to the 'Error Bars' option, choose 'More Error Bars Options...' again and this time choose the Clam Bank. Now add the custom error bars to the graph. When you are finished, the error bars will be on the graph and should look like Figure B.6.

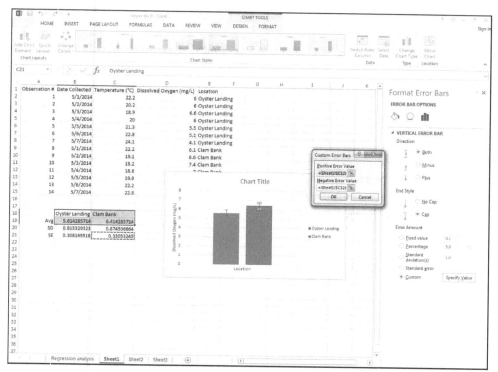

FIGURE B.5

Adding standard error bars to the data figure. Source: Emily G. Baumann. Used with permission from Microsoft Excel.

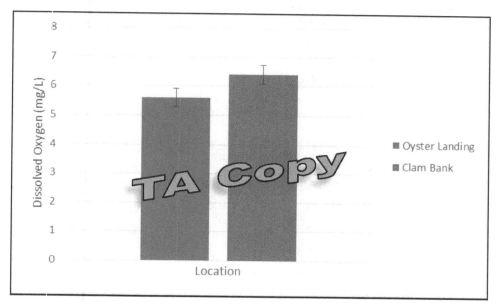

FIGURE B.6

How the figure should appear with error bars added. Source: Emily G. Baumann. Used with permission from Microsoft Excel.

DETERMINING A SIGNIFICANT DIFFERENCE

When looking at the graph you've created, it is hard to tell whether there is really a significant difference between the means. Thus, we need to run a statistical analysis test to determine this.

You can use Excel to perform a *t*-**test** to determine the probability that there is a significant difference between two groups. To perform a "two-tailed" *t*-test implies that the null hypothesis can be rejected by deviations either up or down. This test is determined with the calculation of a p-value. A **p-value** is the probability that the hypothesis being tested is true. A null hypothesis is usually *rejected* if the p-value is < 0.05, and the smaller the p-value the more confident we can be in the conclusions drawn from it.

Let's test the following null hypothesis:

▸ H_0: There is no significant difference between the dissolved oxygen concentrations at Oyster Landing and Clam Bank during the first week of May 2014.

To perform the test in Excel, find a free cell (B23) and type in the equation '=T. TEST(D2:D8,D9:D15,2,2)', where 'array1' is Oyster Landing data and 'array2' is Clam Bank data, and press 'Enter'. Excel will automatically calculate the means and standard deviations to determine the probability of the difference. We are using a standard Type II, two-tailed test, which is indicated by the "2,2" at the end of the equation. A p-value of 0.1021 should be calculated in the cell. Since 0.102 is > 0.05, we will *accept* the null hypothesis to determine that there is *no significant difference* in oxygen concentrations between the two field sites during the first week of May 2014. Don't forget to label the contents of this cell (see Figure B.7).

DETERMINING A SIGNIFICANT RELATIONSHIP

Next we want to know whether there is a relationship between temperature and dissolved oxygen concentration.

Our null hypothesis will be:

▸ H_0: There is no significant relationship between water temperature and the dissolved oxygen concentration and the slope will be zero.

First we want to plot the data to take a look at it. In this case we want to plot temperature vs. dissolved oxygen concentration. Highlight the temperature and dissolved oxygen columns and click on the 'Insert' tab on the toolbar, select 'Scatter', and choose the 'Scatter with only Markers' type (see Figure B.7 on the following page). The graph will appear on the spreadsheet.

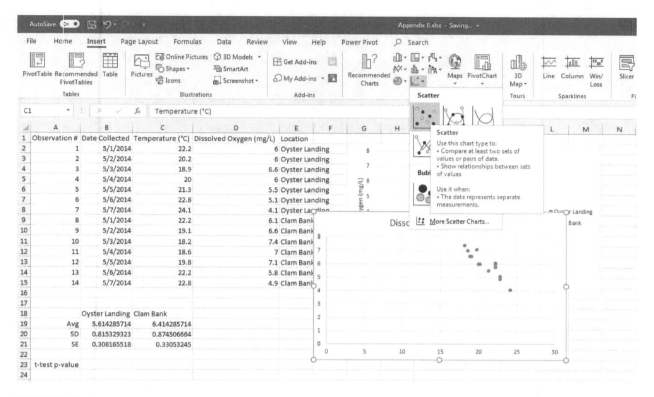

FIGURE B.7

Value of t-test significant difference calculations, and inserting a scatterplot. Source: Emily G. Baumann. Used with permission from Microsoft Excel.

Note that dissolved oxygen is presented as the dependent variable on the Y-axis and temperature is the independent variable on the X-axis. To label the axes (make sure the graph is selected), click the 'Chart Elements' icon, and choose the 'Axis Titles' option. Here you can label the horizontal axis (X-axis) and the vertical axis (Y-axis). Don't forget to include the units! Because we are only plotting one set of data we don't need the title or the legend, so they can be deleted. If you use this graph in a report, a figure caption will be used to explain what the graph is comparing.

Next we will need to determine the equation describing the line and significance of the relationship between the two variables, R^2. To do this we will need to add a **trendline** (best fit line). Under the 'Chart Elements' icon, click on 'Trendline' and follow the arrow to the bottom of the options and click 'More Options...'. Choose the 'Linear' option (see Figure B.8), and check the options for both 'Display Equation on Chart' and 'Display R-squared value on Chart' and close the window. You will see the linear trendline appear on your graph with an equation and R^2 value under it. Your end result should look like Figure B.9.

The R^2 value is 0.80. This tells us that there is a good relationship between the variables and that 80% of the dissolved oxygen variation can be explained by temperature. The slope of the line, −0.44, is not zero so we know that there is a change in dissolved oxygen with a change in temperature, but is the slope significantly different from zero? To answer this question we will have to run another statistical analysis.

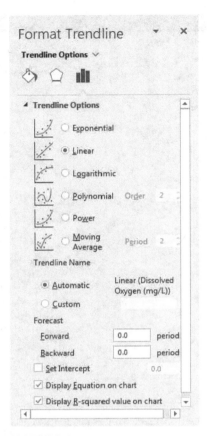

FIGURE B.8

Pop-up window for setting trendline/best fit options. Source: Emily G. Baumann. Used with permission from Microsoft Excel.

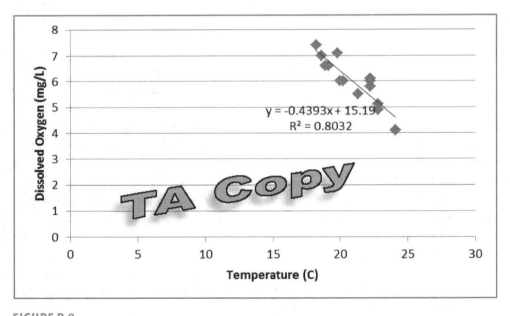

FIGURE B.9

Plotted linear trendline with equation of the line and R^2 value. Source: Emily G. Baumann. Used with permission from Microsoft Excel.

LINEAR REGRESSION ANALYSIS [PC]

To test the hypothesis we will use a Model I linear regression analysis in Excel. Click on the 'Data' tab and then click the 'Data Analysis' icon to the far right of the toolbar. Scroll down the list and select 'Regression' and click 'OK'. The 'Input $\underline{Y}$ Range:' will contain all the values for your dependent variable (dissolved oxygen), so highlight cells D1:D15. Next click on the 'Input $\underline{X}$ Range:' and highlight all the values for your independent variable (temperature), C1:C15 (see Figure B.10). Check the 'Labels' box and the 'Confidence Level: 95%' box, and under the 'Output options' check the 'New Worksheet Ply:' box and type in 'Regression analysis' into the text box and click 'OK'.

A new sheet is created within your spreadsheet containing the 'Summary Output' of your analysis. (If you click on 'Sheet1' you will go back to the sheet containing your data.) When looking at the output, do not get overwhelmed. The value under 'Significance F' (see Figure B.11) is the one important value we are interested in today. This is the p-value for the regression and is given as 1.43679E-05, meaning 0.0000143, so therefore it is < 0.05. The slope is significantly different from zero, so we can *reject* the null hypothesis to conclude that there *is a significant relationship* between dissolved oxygen concentration and water temperature at the two locations in North Inlet during the first week of May 2014.

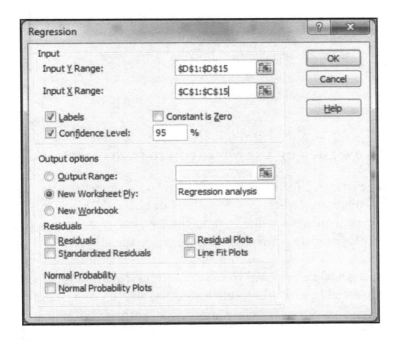

FIGURE B.10

Pop-up window for setting linear regression options. Source: Emily G. Baumann. Used with permission from Microsoft Excel.

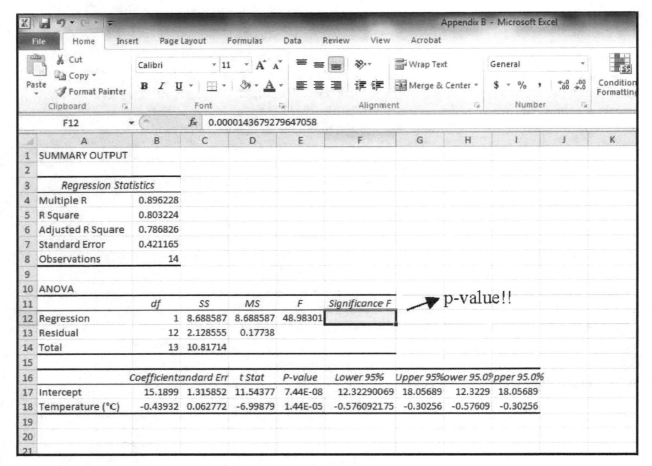

FIGURE B.11
Summary output of a linear regression analysis. Source: Emily G. Baumann. Used with permission from Microsoft Excel.

★ Now click on the Save icon to save your work!! ★

For a list of equations (functions) and how to use them, go to the 'Formulas' tab and click 'More Functions' and choose 'Statistical' or click the 'Help' [?] icon on the toolbar and type 'Functions' under Search. There are too many to describe here and the Microsoft Excel Help search engine is very user-friendly.

Hopefully this guide has provided you with a helpful starting point to get you working more comfortably with Microsoft Excel. You will be using Excel throughout your marine science career to help you organize, analyze, and visualize your data. Remember that the Help Tutorial is always available in the program itself. Furthermore, there are lots of tips and tricks in Microsoft Excel that this guide cannot begin to address. If you do a search on the web for Microsoft Excel Tutorial, numerous websites will come up that will provide you with video tutorials, tips, and tricks for working with Excel. Good luck, and the last piece of advice we can give you is, ALWAYS save your work!

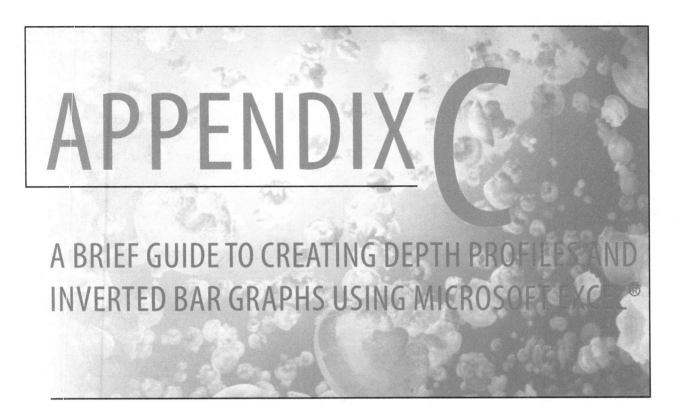

APPENDIX C

A BRIEF GUIDE TO CREATING DEPTH PROFILES AND INVERTED BAR GRAPHS USING MICROSOFT EXCEL®

(Microsoft Excel® 2016 version for Mac and PC)

A scientist's job is to ask questions and then collect data to answer them. Most data that get collected is *quantitative*, and when looking at vast amounts of numbers, it can be difficult to spot trends or make comparisons. A graph is a visual representation of your data that can help you interpret what has been observed. There are numerous ways to visualize your data graphically; this Appendix will guide you in setting up two commonly used graph types in the sciences. Make sure you have already read and understood the steps for analyzing data as described in Appendix B.

INPUT THE DATA

1. On your computer, open a new spreadsheet in Excel and enter the data from the table below. This data was recorded during the noon hour from the dock at Bennett's Point Landing in the ACE Basin of South Carolina. You can find the shortcut for μ by clicking the 'Insert' tab and then choose 'Symbol'— 'More Symbols…', scrolling down until you find it. (Subset: Greek)

Date	Time	Depth (m)	T (°C)	S (PSU)	DO (mg/L)	PAR (μE/m²/s)
3/21/2016	12:34	0	18.8	26.60	6.24	664.90
3/21/2016	12:34	0.5	18.8	26.66	6.13	160.50
3/21/2016	12:34	1	18.8	26.68	6.05	47.45
3/21/2016	12:34	1.5	18.8	26.67	6.09	10.24
3/21/2016	12:34	2.0	18.8	26.70	6.09	2.50
3/21/2016	12:34	2.5	18.9	26.74	6.08	1.20
3/21/2016	12:34	3.0	18.9	26.73	6.09	0.21
3/21/2016	12:34	3.5	18.9	26.69	6.05	0.03
3/21/2016	12:34	4.0	18.9	26.74	6.07	0.00
3/21/2016	12:34	4.5	18.9	26.72	6.06	0.00
3/21/2016	12:34	5.0	18.9	26.71	6.05	0.00

Source: Emily G. Baumann

CREATING A DEPTH PROFILE

In oceanography, scientists create depth profiles to examine patterns exhibited by different parameters of the water column. Such graphs are constructed with the origin (0,0) in the upper left side of the graph so that the surface (depth = 0) is at the top of the graph and *increases* with depth *down* the Y-axis (Figure C.1). However, this is not how the Excel program initially graphs data, so you need to learn how to adjust axes on your graphs.

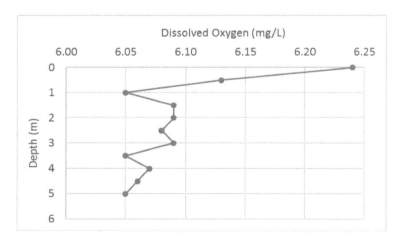

FIGURE C.1

Example of a depth profile of dissolved oxygen (mg/L) in a water column. Source: Emily G. Baumann. Used with permission from Microsoft Excel.

2. Go to the 'Insert' tab and select the 'Scatter with Straight Lines and Markers' option (see Figure C.2). A blank white box will appear on the page.

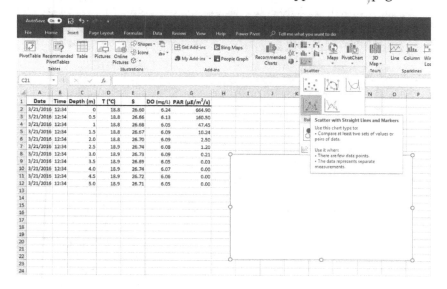

FIGURE C.2

Creating a scatterplot with straight line markers graph by inserting it inside the spreadsheet. Source: Emily G. Baumann. Used with permission from Microsoft Excel.

3. Click on 'Select Data' from the toolbar. In the pop-up window click on 'Add'. A new window will appear with three series of field options. Click the button at the right end of the 'Series X values:' and highlight the data in the salinity column (see Figure C.3). Now press the button again or hit 'Enter'. You are telling Excel that these values are your independent values on the X-axis. Now click the button at the right end of the 'Series Y values:'. Delete the '={1}', highlight the data in the depth column and hit 'Enter'. These values are going to be the dependent values on the Y-axis. Now click 'OK' and 'OK' again.

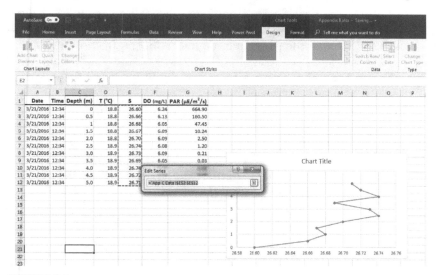

FIGURE C.3

Selecting the data to include in a depth profile of salinity. Source: Emily G. Baumann. Used with permission from Microsoft Excel.

4. The origin (0,0) of your graph is currently in the bottom left corner. For a depth profile, this is not intuitive; the Y-axis needs to be reversed so that depth increases down the Y-axis instead of up. To reverse the Y-axis, right-click on the y-axis and select 'Format axis…', then check the box next to 'Values in reverse order'. This will reverse the order of your Y-axis label (Figure C.4).

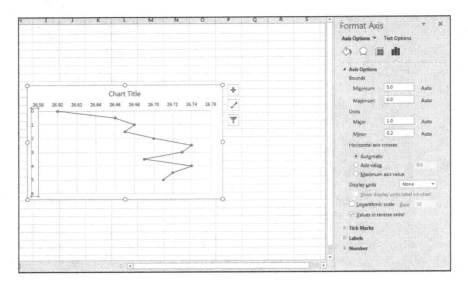

FIGURE C.4

A depth profile of salinity with the Y-axis values reversed to display the surface data at the top of the graph, as it would appear in a cross section of the water column. Source: Emily G. Baumann. Used with permission from Microsoft Excel.

The line displayed by your graph should look like a zig-zag. Excel has automatically set the limits on your X-axis and zoomed into your data set. However, the salinity is not actually as dramatic as it first appears. Let's modify the X-axis to get a more realistic look at salinity throughout the water column.

5. Right-click on the X-axis and go to "Format axis…". Change the Minimum bound to "0.0" and the Maximum bound to "30".

6. Label the axes and don't forget to include the units!! Your resulting graph should look like the one below (Figure C.5).

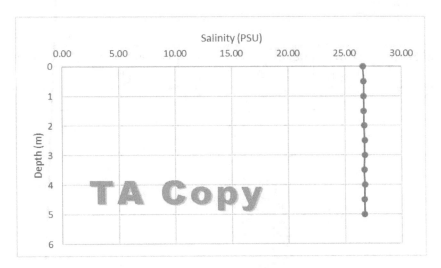

FIGURE C.5
A depth profile of salinity (PSU). Source: Emily G. Baumann. Used with permission from Microsoft Excel.

As you can see, the resulting profile of salinity actually exhibits relative stability throughout the water column. Practice with making more depth profiles using the data provided for temperature, dissolved oxygen, and light attenuation (PAR).

CREATING AN INVERTED BAR GRAPH

Marine biologists who study trophic levels often prefer to examine their results using an inverted stacked bar graph as in the example below (Figure C.6). For this graph, data was collected for five different trophic levels during two tidal stages (high tide and low tide). The measurements are presented as biomass in milligrams of carbon per cubic meter.

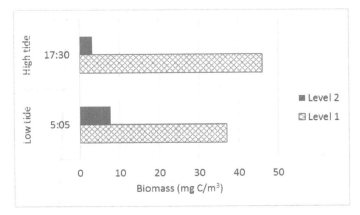

FIGURE C.6
Biomass (mg C/m^3) comparison of 2 trophic levels collected at two tidal stages (high and low tide). Source: Emily G. Baumann. Used with permission from Microsoft Excel.

7. To make this graph in Excel, enter the following data table into an Excel spreadsheet (Figure C.7). Data presented in the table below are biomass values in units of mg C/m^3.

Tidal Stage	Time	Level 1	Level 2	Level 2.5	Level 3	Level 3.5
Low tide	5:05	37	7.7	3.02	1.42	0.19
High tide	17:30	46	3.2	25.10	4	0.3

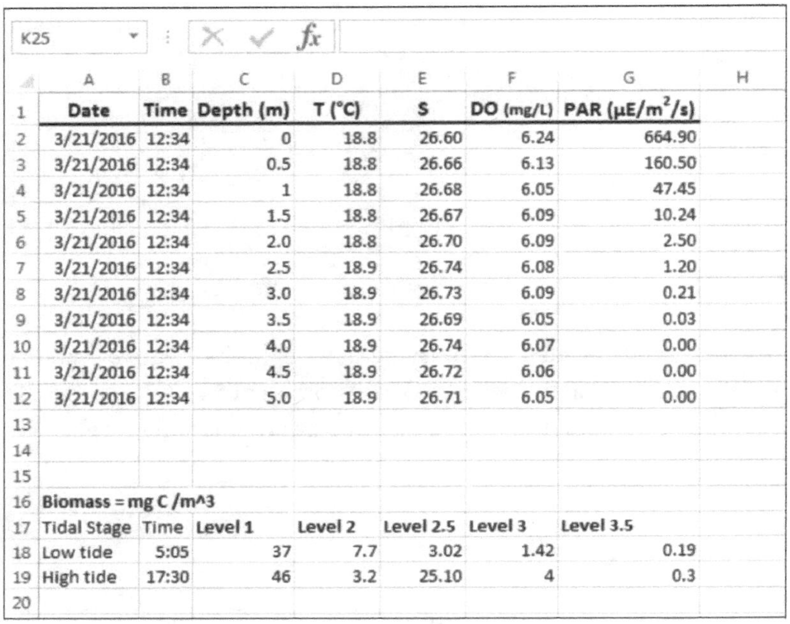

	A	B	C	D	E	F	G	H
1	Date	Time	Depth (m)	T (°C)	S	DO (mg/L)	PAR (µE/m²/s)	
2	3/21/2016	12:34	0	18.8	26.60	6.24	664.90	
3	3/21/2016	12:34	0.5	18.8	26.66	6.13	160.50	
4	3/21/2016	12:34	1	18.8	26.68	6.05	47.45	
5	3/21/2016	12:34	1.5	18.8	26.67	6.09	10.24	
6	3/21/2016	12:34	2.0	18.8	26.70	6.09	2.50	
7	3/21/2016	12:34	2.5	18.9	26.74	6.08	1.20	
8	3/21/2016	12:34	3.0	18.9	26.73	6.09	0.21	
9	3/21/2016	12:34	3.5	18.9	26.69	6.05	0.03	
10	3/21/2016	12:34	4.0	18.9	26.74	6.07	0.00	
11	3/21/2016	12:34	4.5	18.9	26.72	6.06	0.00	
12	3/21/2016	12:34	5.0	18.9	26.71	6.05	0.00	
13								
14								
15								
16	Biomass = mg C /m^3							
17	Tidal Stage	Time	Level 1		Level 2	Level 2.5	Level 3	Level 3.5
18	Low tide	5:05	37		7.7	3.02	1.42	0.19
19	High tide	17:30	46		3.2	25.10	4	0.3
20								

FIGURE C.7

The addition of biomass data to the spreadsheet. This data will be used to make an inverted bar graph.
Source: Emily G. Baumann. Used with permission from Microsoft Excel.

8. In Excel, select the entire data table (tide stages, times, and levels) and click the 'Insert' tab. Go to the bar chart icon and select the 2-D Clustered Bar option (Figure C.8).

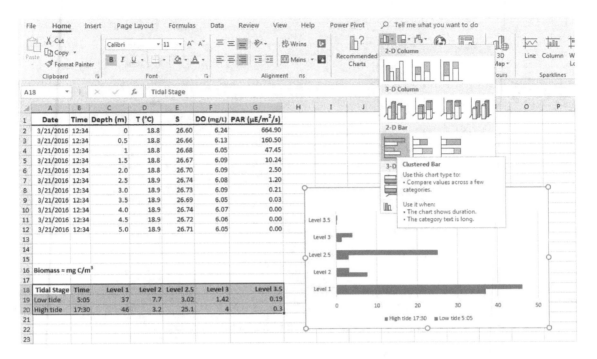

FIGURE C.8

Inserting a clustered bar graph using the desired biomass data set. Source: Emily G. Baumann. Used with permission from Microsoft Excel.

9. Click on 'Switch Row/Column' under the 'Design' tab to tell Excel to switch the way it is graphing the data. You want to graph the data grouped by Row instead of by Column. Your bar graph should look like the example in Figure C.9.

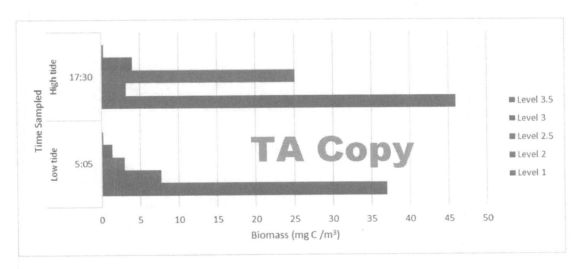

FIGURE C.9

Comparison of biomass (mg C/m³) for five trophic levels measured during two tidal stages (high and low tide).

Source: Emily G. Baumann. Used with permission from Microsoft Excel.

10. To add the superscript to your axis label, highlight the 3 within your axis title, right-click on it and choose 'Font'. Check the box for Superscript and click OK.

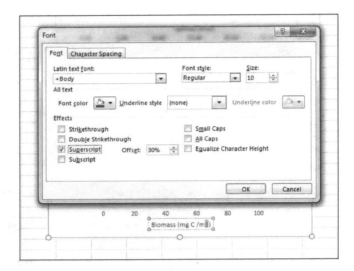

FIGURE C.10

Creating superscript text in an axis label on a graph. Source: Emily G. Baumann.
Used with permission from Microsoft Excel.

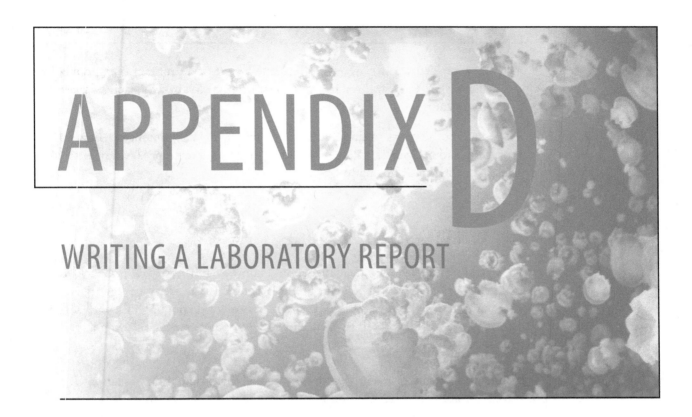

APPENDIX D

WRITING A LABORATORY REPORT

Scientific writing is a style of writing that is very different from a typical English paper or history report. The style of a lab report or scientific paper should be objective, free of personal bias, and without embellishments or flowery language. It is never written in a fictional or storybook style, or other types of prose. Its purpose is to describe to a scientific audience exactly how an experiment was done, the results that were obtained, and what the author thinks the results mean. Scientific reports never use bullets or outlines; while it should be direct and descriptive, a report should also be flowing and easy to read. Every lab report and scientific paper should include the following sections described below.

I. INTRODUCTION

An introduction provides the reader with the purpose of the experiment. You should introduce the subject of your experiment and provide a brief description of research done by previous investigators and how your experiment ties into this background. A conceptual background should also be included, describing the mechanics, processes, and interrelationships that relate to your experiment. Place the questions you are asking in a broader context.

    A. Introduce the problem.

    B. Define the questions, hypotheses, or purpose of the experiment.

    C. Why is this investigation important?

D. If you are investigating more than one aspect of a problem (i.e., population dispersion, energy, diversity) you must briefly explain why each of these are important.

E. What other work in this field has been done? Include references—how does it relate to the general field you are studying? Good references will help provide you with this information.

II. MATERIALS AND METHODS

After reading this section, the reader should know exactly how you have tackled the problem. Each step should be described in order, so that a later investigator can utilize your procedure to repeat the experiment. However, do not use bullet or list form; describe how you performed the experiment in complete sentences using paragraph structure.

Included in this section is a description of the environment and location. If geographic or climate variability might influence your results, then the source of your material and/or source location, along with a description of prevailing environmental conditions should be included. If seasonal or diurnal variation is of importance, reference to the time of the experiment is necessary. Other scientists will need to know how comparable an experiment would be that they model after yours, particularly if it is affected by environmental conditions.

When reporting the name of an organism, the first time you mention the name, always give the full scientific name of organisms used in the format of: *Genus species*. Every subsequent mention of the name for that species, and any species belonging to that genus, should then be abbreviated as: *G. species*.

A. Precisely describe your experiment so that a later worker can repeat it in the same way. You may assume the reader knows something about the subject (*e.g.,* You don't have to say "Cut *Spartina* was placed in a can and dried in an oven." Rather, "A square meter sample of *Spartina* was dried in an oven and dry weight was measured." Here you can assume the reader will know you used a scale to measure weight.)

B. Make sure you include important information such as:

1. Field Information
   a. Where your samples came from
   b. When they were taken
   c. Method of collection, sample sizes and numbers collected, categories or types of observations made.

2. Lab Procedures
   a. Go step by step. Write it in a descriptive manner, not bulleted or outlined.

    b.   You can use the procedure instructions from lab handouts as a guide but do not include the exact phrasing in your procedures—put it in your own words.

    c.   Be clear about what measurements were used for each of the parameters.

3. Computations and Statistics

    a.   If statistics are used, be clear about what statistics were used for each of the parameters.

    b.   Include statistical formulas used.

4. Style

    a.   Use third person point-of-view or passive voice at all times.

    b.   DO NOT use first person pronouns (*I, we, you, me, our*, etc.)

    c.   'Datum' is singular, 'data' is plural. 'Genus' is singular, 'genera' is plural. Make sure your verbs agree with these nouns.

III. RESULTS

The purpose of the Results section is simply to report what results were obtained. If you describe your data in the text, they should be stated simply and objectively. If the data are given in the form of tables, charts, or figures (graphs, illustrations, etc.), call attention to each and summarize the data given there. You cannot include a table or figure without referring to it in the text. Tables or figures should always be numbered in the order they are referred to in the text. No explanations of results should be provided here, nor should you discuss reasons for your data or experimental results. Just state the results, leaving explanations, comparisons with other data, etc., to the Discussion section.

    A.   Report your results. Do not explain or give reasons for trends in the Results section.

    B.   Try to always use tables and graphs to provide the reader with an easily viewed representation of your data. A picture is worth a thousand words!

    C.   Describe general trends that you see in the data. This is not part of the Discussion, but be aware that there is a fine line between the Results and Discussion—just describe what is seen, not what it means or indicates.

    D.   When you use tables and figures to summarize data, make sure you label them correctly with a caption.

    E.   Never provide a table or graph without an explanation in the text referring the reader to this visual representation of your data. For example, if you have a Table 1, then "Table 1 contains density measurements for 10 beach samples" must be included within the text of the Results section.

IV. DISCUSSION

This is perhaps the most important section of your report or paper since it (hopefully) will make sense of the data reported in the Results section. In the Discussion, you will interpret your data, provide explanations for your results, and/or consider alternative hypotheses. If the experiment has told you anything pertinent to the subject and background as discussed in the Introduction, it should be pointed out.

A.   Discuss the processes that produced the features you saw, how these processes relate to those features or contribute to the results you obtained.

B.   The Discussion should include references. How do your trends compare to those in other studies? Compare your results to those found by other investigators, and explain differences or similarities seen.

C.   Interpret results in the context of a broader picture, and give reasons for trends seen. Always refer directly to your data.

D.   When drawing a conclusion based on your data, you must discuss and consider all of your data, not just those which support your views. Discuss uncertain findings.

E.   Focus on carefully analyzing and interpreting your results. Do not concentrate on searching for the "right" answer or results that match those found by other groups.

V. SUMMARY (OR) CONCLUSION

You may use either a summary or conclusion format, depending upon which better serves your report or paper. If a summary is used, conclusions should be drawn in the Discussion section. The differences in the formats are presented below.

A.   In a Summary section, the salient points of each section, except the Introduction, are repeated in a sentence or two. They are usually numbered, as follows:

    1.   What was done
    2.   What happened
    3.   What it means

No new information is presented in a Summary. Everything has been said in previous sections and is simply and briefly reiterated in the Summary.

B.   In a Conclusion section (preferred method), the interpretations presented in the Discussion should be used to form an overall conclusion about your experiment. In other words, you should "pull it all together" in a Conclusion. Wrap it up in this final paragraph.

VI. LITERATURE CITED

Every academic discipline has its own individual method of citing reference material. Scientific literature citations probably differ most from the methods you have been taught to use, but it is the most concise and you should get used to using it.

A. Citations

In the *body* of your paper, you must cite the author of any previous work or the source of any information to which you refer. This includes numbers, formulas, descriptions you obtained from other sources, etc. However, you do not footnote at the bottom of the page or at the end of the paper as you do with English papers. You simply give parenthetically the author's name and the year the paper was published at the end of your sentence. Or, you may make the author's name a part of your sentence by giving the date immediately after it in parentheses. If the date of the work is important, this too may be incorporated into the sentence.

Here are some examples:

1.   There is a considerable delay in maturation of *Crassostrea virginica* living under conditions of seriously depressed salinity (Butler, 1949).

2.   Raymont (1963) claims that most marine zooplankton display diurnal vertical migrations.

3.   It was not until 1953 that the double helix structure for DNA became popularized by Watson and Crick.

B. Literature Cited (or) References

At the end of your paper, in a separate section, all of the literature cited (and *only* those references cited in the text) are listed in alphabetical order. Papers by a single author are listed before papers by that author collaborating with someone else. If you have several papers with the same first author and different additional authors, the papers are listed alphabetically starting with the second author (or third, if necessary). Papers by identical authors are listed chronologically. Below are some examples showing the style to be used for different types of references:

Bainbridge, R. 1961. "Migrations." Chap. 12. In: *The Physiology of Crustacea*, Vol. 2, pp. 431–463, ed. by T. H. Waterman. Academic Press, New York.
Butler, P. A. 1949. Gametogenesis in the oyster under conditions of depressed salinity. *Biol. Bull.* 96: 263–269.
Raymont, J. E. G. 1963. *Plankton and Productivity in the Oceans*. Pergamon Press, Oxford. 660 pp.

Watson, J. D. 1963. The involvement of RNA in the synthesis of proteins. *Science* 140: 17–26.

Watson, J. D. 1970. *Molecular Biology of the Gene*. W. A. Benjamin, Inc., New York, 662 pp.

Watson, J. D. and F. C. Crick. 1953. Molecular structure of nucleic acids. *Nature* 171: 737–738.

C. Reference Formatting

There are multiple ways of formatting your References. It is best to follow a well-known journal style or handbook, and be consistent throughout. Always italicize (or underline) titles of books and titles of journals. Do not italicize or underline article titles or chapter headings. If you need to cite a type of reference not listed above, such as a website, ask your lab instructor for assistance or visit the Thomas Cooper Library website using the following link: http://www.library.sc.edu. This link provides several websites with detailed instructions on formatting citations.

FORMATTING YOUR PAPER OR REPORT

Any paper or lab report should *always* use the formatting listed below. Your instructor must approve any variations from this formatting beforehand.

1. Title page, with title of report, name, section, and any other information required by your instructor.
2. 12-point font
3. 1-inch margins on all sides
4. Double-spaced paragraphs
5. Indent new paragraphs a minimum of five spaces
6. Set headings apart from body text
7. Label figures, tables, and diagrams (including axes)
8. Page numbers

Do not wait until the last minute to write your lab report! First, make a detailed outline and write a draft of your report. Let it sit for a day or so, re-read it for clarity and make corrections, then write the final copy. Have a friend read your first draft and make suggestions—you will almost always miss a mistake. If you are not good at grammar, it is very important to get someone else to edit your work.

Remember, writing a lab reports or scientific paper is a skill that is learned through repetition, and the more you practice at it, the better you will become!

APPENDIX E

CONVERSIONS REVIEW AND WORKSHEET

DIMENSIONS AND UNITS

In general, scientists measure a few fundamental **dimensions**, or measurements, from which all other variables can be derived: time (T), length (L), mass (M), and temperature. From combinations or derivations of these basic units, everything else we are interested in can be determined. For example:

Density: measured in grams (*mass*) per cubic meter (*length*)

Energy: measured in Joules (J). One Joule is the amount of work done when a force of 1 Newton moves over a *length* of 1 meter. A Newton is the force required to give a *mass* of 1 kilogram an acceleration of 1 meter per second (*length* and *time*).

Pressure: measured in millibars, bars, or Pascals (100,000 bars). One Pascal (Pa) is the pressure generated by a force of 1 Newton acting on an area of 1 square meter.

As you can see from these examples, for scientists to be able to express their measurements, they must use **units**, a defined quantity adopted as the standard for that type of measurement. We use units constantly: seconds, inches, pounds, tons, meters, etc. There are several different systems that have been developed to express dimensions in different units. The most common system, the **Systeme International** (SI for short), is based on the units of meter-kilogram-second. Another system is the English system, in which units of the foot-pound-second are used for length, mass, and time, respectively. The SI system is used by the vast majority of the sciences.

The **Metric System** is another unit system based on powers of 10. Its fundamental dimensions are measured in meters, grams, and seconds. It also includes prefixes to designate multiples of ten:

Prefix	Symbol	Factor	Decimal
pico	p	10^{-12}	0.000000000001
nano	n	10^{-9}	0.000000001
micro	μ	10^{-6}	0.000001
milli	m	10^{-3}	0.001
centi	c	10^{-2}	0.01
deci	d	10^{-1}	0.1
deka	da	10	10
hector	h	10^{2}	100
kilo	k	10^{3}	1,000
mega	M	10^{6}	1,000,000
giga	G	10^{9}	1,000,000,000
tera	T	10^{12}	1,000,000,000,000

When dealing with very large or very small values, it's often easier to use prefixes with these units to designate a multiple of 10. For example, 1,000 meters is 1 kilometer, 1,000,000 nanometers is 1 millimeter, etc.

CONVERSIONS: CHANGING UNITS

In general, Americans are most familiar with the English system, while oceanographers, marine biologists, geologists, and the vast majority of scientists use the metric system. It is essential that students of any scientific discipline know not only how to convert between these systems, but between different units in general. Why? Often, you can work through a difficult calculation by just converting from one unit to another to get to the answer you need. To do conversions successfully, there are a few principles you need to keep in mind.

1. Dividing a unit by itself (in any form) is equal to 1.

When converting units, keep in mind the principle that anything multiplied by 1 equals itself, and so conversely, anything divided by itself is equal to 1. Thus, 12 inches divided by 12 inches equals 1. Similarly, 12 inches divided by 1 foot equals 1, as does 1 inch divided by 2.54 centimeters, 1.852 kilometers divided by 1 nautical mile, etc.—you get the idea. Conversions are an essential method to obtain a new unit without actually altering the original *value* of the measurement. You may change the numbers, but 1 foot is still 12 inches in length no matter how you look at it.

When students attempt to do math problems, many times the difficult part (and where most of the points are lost) is not in the concept, but in the conversions. Let's look at how you can set up these conversions and avoid mistakes:

Example 1: Convert 3 nautical miles to meters.

$$3 \; \cancel{\text{n mi}} \; \times \; \frac{1.852 \; \cancel{\text{km}}}{1 \; \cancel{\text{n mi}}} \; \times \; \frac{1000 \; \text{m}}{1 \; \cancel{\text{km}}} \; = \; 5{,}556 \; \text{m}$$

Unit conversion pair

Where a student might miss this conversion is not placing the kilometer unit on top in the first unit conversion pair, or meters on top in the second. This leads us to our next principle.

2. Set up unit pairs so that you can cancel out all but the final answer's unit.

This principle seems obvious, but many mathematical mistakes are done when students forget they can put either of the unit pair in the numerator (top) or the denominator (bottom). Look at the example below:

Example 2: Convert 3,000 meters to nautical miles.

$$3{,}000 \; \cancel{\text{m}} \; \times \; \frac{1 \; \cancel{\text{km}}}{1{,}000 \; \cancel{\text{m}}} \; \times \; \frac{1 \; \text{n mi}}{1.852 \; \cancel{\text{km}}} \; = \; 1.62 \; \text{n mi}$$

No matter what unit you are trying to obtain, always set up your pairs so that the numerator unit of one pair can cancel out an identical unit in the denominator for another pair. As long as each unit pair is equal to 1, determining which one is in the numerator vs. the denominator is solely a function of how you want the units to cancel. That's really all there is to conversions.

Here are a few more involved examples:

Example 3: You dissolve 35 grams of salt in 1L of freshwater. What is the density in g/cm^3?

$$\frac{35 \text{ g}}{1 \text{ L}} \times \frac{1 \text{ L}}{1{,}000 \text{ mL}} \times \frac{1 \text{ mL}}{1 \text{ cm}^3} = 0.035 \text{ g/cm}^3$$

Example 4: How do you convert 15 µmols/L to µmols/m^3?

$$\frac{15 \text{ µmols}}{\text{L}} \times \frac{1 \text{ L}}{1{,}000 \text{ ml}} \times \frac{1 \text{ ml}}{1 \text{ cm}^3} \times \frac{(100 \text{ cm})^3}{(1 \text{ m})^3} = 15{,}000 \text{ µmol/m}^3$$

Example 5: Convert 15 µmols/L to µmols/kg.

$$\frac{15 \text{ µmols}}{\text{L}} \times \frac{1 \text{ L}}{1{,}000 \text{ ml}} \times \frac{1 \text{ ml}}{1 \text{ cm}^3} \times \frac{1 \text{ cm}^3}{\text{g}} \times \frac{1{,}000 \text{ g}}{\text{kg}} = 15 \text{ µmol/kg}$$

Note in this example that you need to include the density of freshwater (1 g/cm^3) in order to get to the units you need (kg) and cancel those you didn't (L). You can also use seawater density if you are given a density value, or the weight of salts in a volume of seawater as in Example 3. This is a very helpful trick to remember when converting volumes to weights.

Example 6: Convert 23 km/hr to cm/sec.

$$\frac{23 \text{ km}}{\text{hr}} \times \frac{1 \text{ hr}}{60 \text{ min}} \times \frac{1 \text{ min}}{60 \text{ sec}} \times \frac{1{,}000 \text{ m}}{1 \text{ km}} \times \frac{100 \text{ cm}}{1 \text{ m}} = 639 \text{ cm/sec}$$

SOME IMPORTANT VARIABLES IN MARINE SCIENCE

1. **Area:** Area is a squared length (L^2).

2. **Volume:** Volume is a cubic length (L^3). A handy conversion to memorize is that there are 1,000 liters of water per cubic meter (m^3).

3. **Density:** Density is equal to a mass divided by a volume (M/L^3). In SI units it is most often expressed in kg/m^3, but you will often see it in g/cm^3 as well.

4. **Velocity:** Also known as a rate, or speed, and is measured as a unit length over a unit time (L/T).

CONVERSION FACTORS FOR SOME COMMON UNITS

Length	
1 inch (in)	= 2.54 centimeters (cm)
1 foot (ft)	= 0.3048 meters (m)
1 meter (m)	= 100 cm = 1,000 mm
1 cm	= 10 millimeters (mm)
1 kilometer (km)	= 1,000 m = 100,000 cm
1 yard (yd)	= 0.91 m
1 statute mile (st mi)	= 1.609 km = 5,280 ft
1 nautical mile (n mi)	= 1.852 km = 1.15 st mi
1 knot	= 1 n mi / hour (it is a velocity)
1 fathom	= 1.829 m = 6 ft
Volume	
1 m^3	= 1,000 liters (L) = 1.307 yd^3
1 gallon (gal)	= 4 quarts (qt) = 3.785 liters
1 cubic centimeter (cc or cm^3)	= 1 milliliter (ml)
Mass	
1 ounce (oz)	= 28.35 grams (g)
1 pound (lb)	= 16 oz = 453.6 g
1 kilogram	= 2.205 lb
1 metric ton	= 2,205 lb = 1,000 kilograms (kg)

PRACTICE PROBLEM SET

Directions

Your instructor will assign you portions or all of these questions as homework. These questions are slightly more difficult than the problem set provided in the MSCI 101 Lab Manual, so if you are uncertain about doing conversion problems it is recommended that you work through the MSCI 101 Practice Problem Set first (your instructor will provide you with a copy). Some of these problems will require you to look up information in your textbook or on the web. All the conversions you need are provided in the Conversion Factors Table. Write your answers out on a separate piece of paper and *always show your work*—partial credit can be given if your instructor can see where you were headed with your thinking!

1. A cruise ship is traveling down the coast from Charleston, SC to Miami, FL (~600 miles). If the ship is going 10 knots, how many days will it take to complete?

2. The state fish of South Carolina is the striped bass. The largest caught by rod and reel is 78 lbs, 8 ounces. How much is that in kilograms?

3. Put the following distances in order from shortest to longest:

   100 km, 100 n mi, 75 miles, 400,000 ft

4. Put the following volumes in order from smallest to largest:

   1,000 gal, 1,000 L, 2 yd^3, 1.5 m^3

5. A research vessel takes 5 days to travel the 1,600 miles from Miami, FL to Boston, MA onboard a ship. How fast is the ship going in knots?

6. The size of your ship's hold is 10 ft × 10 ft × 20 ft. The average fish size is 18 cm × 12 cm × 5 cm.

   a. How many fish can the ship hold?

   b. Each fish weighs 3 lbs. How many metric tons does the ship hold?

   c. What is the average density of the fish in g/cm^3?

7. A local fishery is allowed to stock fish in a 1 mi^2 area of water off the coast of Myrtle Beach. To stock the correct amount of fish, the fishery must first determine how big the location is. Assuming an average water depth of 7 fathoms, what is the volume of water that the fishery needs to stock:

   a. in L?

   b. in m^3?

8. Seawater has an average particle load of 50 mg/L. How many gallons would you have to filter in order to obtain 0.5 lbs of material?

LAB SOURCES

LAB 1: THE PLANKTON

Castro, P. and Huber, M.E. 2000. *Marine Biology*. 3rd Ed. McGraw Hill: Boston.

Lalli, C.M. and Parsons, T.R. 1997. *Biological Oceanography: An Introduction*. 2nd Ed. Butterworth-Heinemann: The Open University.

Marine Science Program, University of South Carolina. "The Present Day Marine Environment Laboratory Manual," MSCI 112 Spring 2002.

Pinet, P.R. 2000. *Invitation to Oceanography*. 2nd Edition. Jones and Bartlett Publishers: Sudbury, Mass.

LAB 2: METHODS OF CHLOROPHYLL ANALYSIS

Boucher, J. and Sahl, L.E. 2006. An Introduction to Finding Context. *Oceanography* 19(3): 146–149.

> The overall concept of the "Chlorophyll Analysis" activity is very generally based on the lab exercise called "An Introduction to Finding Context," by J. Boucher and L. E. Sahl (2006). In this activity, the authors incorporate a standard method of chlorophyll measurement to introduce scientific concepts, methodology, and data interpretation to students.

Castro, P. and Huber, M. E. 2000. *Marine Biology*. 3rd Ed. McGraw Hill: Boston.

ChevronTexaco and UCSC Remote Sensing Group. 2001. "Remote Sensing for Environmental Applications." Accessed 12/4/07. http://www.es.ucsc.edu/~hyperwww/chevron/.

Feldman, G. SeaWiFS Project Homepage. Accessed 12/4/07. http://seawifs.gsfc.nasa.gov/SEAWIFS.html.

"MODIS Specifications." MODIS Web. NASA. Accessed 08/19/2014. http://modis.gsfc.nasa.gov/about/

Pagani, M., Arthur, M. A., and Guber, A. L. 1998. *The Sea Around Us: Laboratory Manual*. 2nd Ed. Kendall/Hunt Publishing Co.: Dubuque, IA.

Pinet, P. R. 2000. *Invitation to Oceanography*. 2nd Ed. Jones and Bartlett Publishers: Sudbury, Mass.

Short, N. M. 2003. Remote Sensing Tutorial. Accessed 12/4/07. http://rst.gsfc.nasa.gov/Homepage/Homepage.html.

LAB 3: PRIMARY PRODUCTION

Broecker, W. S. and Peng, T. H. 1982. *Tracers in the Sea*. Lamont Doherty Geological Observatory, Columbia University, New York.

Garrison, T. 2002. *Oceanography: An Invitation to Marine Science*. 4th Ed. Thomson Learning, Inc.: California.

Hardee, M. L. and Benitez-Nelson, C. 2004. *Introduction to Oceanography Lab Manual*. Kendall/Hunt Publishing Co.: Dubuque, IA.

Libes, S. M. 1992. *An Introduction to Marine Biogeochemistry*. John Wiley & Sons, Inc.: New York.

Marine Science Program, University of South Carolina. "The Present Day Marine Environment Laboratory Manual," MSCI 112 Spring 2002.

Pinet, P. R. 2000. *Invitation to Oceanography*. 2nd Ed. Jones and Bartlett Publishers: Sudbury, Mass.

LAB 4: EXPERIMENTAL AND STATISTICAL ANALYSES

Dytham, C. 2003. *Choosing and using statistics: a biologist's guide*. 2nd Ed. Blackwell Publishing Co.

Marine Science Program, University of South Carolina. "The Present Day Marine Environment Laboratory Manual," MSCI 112 Spring 2002.

Phillips, Jr., J. L. 1992. *How to Think About Statistics*. Revised Ed. W.H. Freeman & Co.

Sokal, R. R. and Rohlf, F. J. 1994. *Biometry: The Principles and Practice of Statistics in Biological Research*. 3rd Ed. W H Freeman & Co.

"The Assateague Naturalist." 2001. Accessed 12/4/2007. http://www.assateague.com/shells.html.

LAB 5: THE "NUTS AND BOLTS" OF TAXONOMY

BioEd Online: Biology Teacher Resources from Baylor College of Medicine. 2006. Accessed 12/4/07. http://www.bioedonline.org/slides/slide01.cfm?q=%22phylogenetic%22

Brusca, R. C. and Brusca, G.J. 1990. *Invertebrates*. Sunderland, Mass.: Sinauer Associates, Inc.

Coppard, S. 2006. "International Commission on Zoological Nomenclature." Accessed 12/4/07. http://www.iczn.org/

Cummins, R. H. 1996. "The 'Nuts and Bolts' of Taxonomy and Classification." Accessed 12/4/07. http://jrscience.wcp.muohio.edu/lab/TaxonomyLab.html

Marine Science Program, University of South Carolina. "The Present Day Marine Environment Laboratory Manual," MSCI 112 Spring 2002.

Parrish, J. 2005. "Dichotomous Keys Lab." Accessed 12/4/07. http://fish.washington.edu/classes/fish250/.

Wiley, E. O. 1981. *Phylogenetics: The Theory and Practice of Phylogenetic Systematics*. New York: John Wiley & Sons, Inc.

LAB 6: FOOD WEBS AND TROPHIC LEVELS

Belle W. Baruch Institute for Marine and Coastal Sciences, Baruch Marine Field Laboratory. 2004. North Inlet—Winyah Bay National Estuarine Research Reserve.

Brantley, S. Stone, S., and Young, R. 2001. The Dolphins of North Inlet: A Rising Tide Project. "Photo-Identification Activity" and "The Ecological Role of Bottlenose Dolphins, *Tursiops truncatus*, in an Estuarine Food Web."

> This activity was developed by Steven Brantley, an undergraduate marine science major at Coastal Carolina University, and Shannon Stone, a high school science teacher at Socastee High School in Horry County, SC. It is based on their research with Dr. Rob Young, a marine science professor at Coastal Carolina University. The activity was reviewed by Dr. Young and revised after being tested in the classroom. All photos were taken by R. Young and collected under NOAA-NMFS permit 976-1582.

Castro, P. and Huber, M. E. 2000. *Marine Biology*. 3rd Ed. McGraw Hill: Boston.

Lalli, C. M. and Parsons, T. R. 1997. *Biological Oceanography: An Introduction*. 2nd Ed. Butterworth-Heinemann: The Open University.

Trefil, J. and Hazen, R. M. 2001. *The Sciences: An Integrated Approach*. 3rd Ed. John Wiley & Sons, Inc.: New York.

Young, R. F. and H. D. Phillips. 2002. Primary production required to support bottlenose dolphins in a salt marsh creek system. *Marine Mammal Science* 18(2): 358-373.

LAB 7: SURVEY OF MARINE ORGANISMS I AND II

Brusca, R. C. and G. J. Brusca. 1990. *Invertebrates*. Sunderland, MA: Sinauer Associates, Inc.

Campbell, N. A. and Reece, J. B. 2002. *Biology*, 6th Ed. Menlo Park, CA: Benjamin/Cummings Publishing Co., Inc.

Maddison, D. R. 2001. The Tree of Life Web Project. http://www.tolweb.org.

Marine Science Program, University of South Carolina. "The Present Day Marine Environment Laboratory Manual," MSCI 112 Spring 2002.

Scholtens, B. 1996. *Biology 112 Laboratory Manual*, 2nd Ed. Burgess Publishing.

LAB 8: ADAPTATIONS OF MARINE ORGANISMS

Berkowitz, S. 2001. "Introduction to Marine Science: MSCI 111L Laboratory Manual." Coastal Carolina University.

Castro, P. and Huber, M.E. 2000. *Marine Biology*. 3rd Ed. McGraw Hill: Boston.

Marine Science Program, University of South Carolina. "The Present Day Marine Environment Laboratory Manual," MSCI 112 Spring 2002.

Pinet, P. R. 2000. *Invitation to Oceanography*. 2nd Ed. Jones and Bartlett Publishers: Sudbury, Mass.

LAB 9: DEEP OCEAN ECOSYSTEMS

Castro, P. and Huber, M. E. 2000. *Marine Biology*. 3rd Ed. McGraw Hill: Boston.

Lalli, C. M. and Parsons, T. R. 1997. *Biological Oceanography: An Introduction*. 2nd Ed. Butterworth-Heinemann: The Open University.

NeMO Education material. Accessed 12/4/07. http://www.pmel.noaa.gov/vents/nemo1998/curriculum.html.

Pagani, M., Arthur, M. A., and Guber, A. L. 1998. *The Sea Around Us: Laboratory Manual*. 2nd Ed. Kendall/Hunt Publishing Co.: Dubuque, IA.

Pinet, P. R. 2000. *Invitation to Oceanography*. 2nd Ed. Jones and Bartlett Publishers: Sudbury, Mass.

Pipkin, B. W., Gorsline, D. S., Casey, R.E., Dunn, D. A., and Schellenberg, S.A. 2001. *Laboratory Exercises in Oceanography*. 3rd Ed. W.H. Freeman and Co.: New York.

Project NeMO: New Millennium Observatory. NOAA Pacific Marine Environmental Laboratory Vents Program. Accessed 12/4/07. http://www.pmel.noaa.gov/vents/nemo/.

Yancey, P. H. "DEEP-SEA Pages: Biological Research and Information on Deep-Sea Habitats and Adaptations." Accessed 12/4/07. http://people.whitman.edu/~yancey/deepsea.html

APPENDIX A: HOW TO USE A MICROSCOPE

Lazaroff, M. "How to Use a Microscope Properly!" Accessed 12/4/07. http://shs.westport.k12.ct.us/mjvl/biology/microscope/microscope.htm

Marine Science Program, University of South Carolina. "Evolution of the Marine Environment Laboratory Manual," MSCI 111 Fall 2002.

"Microscopes: Parts Self-Test." Accessed 12/4/07. http://www.southwestschools.org/jsfaculty/Microscopes/index.html.

Scholtens, B. 1995. *Biology 111 Laboratory Manual*. Burgess Publishing Co.: Edina, MN.